100 ADS Design Examples

Based on the Textbook:

RF and Microwave Circuit Design

A Design Approach Using **(ADS)**

100 ADS Design Examples

Based on the Textbook:

RF and Microwave Circuit Design

A Design Approach Using **(ADS)**

Ali A. Behagi

Techno Search

Ladera Ranch, CA 92694

100 ADS Design Examples

Based on the Textbook:

RF and Microwave Circuit Design

A design Approach using (**ADS**)

Copyright © 2016 by Ali A. Behagi
ISBN 978-0-996-4466-2-4

First Published, January 2016

Published in USA

Techno Search

Ladera Ranch, CA 92694

Foreword

100 ADS Design Examples, based on the author's *RF and Microwave Circuit Design* textbook is a hands-on step-by-step RF and microwave circuit design examples for university students and a valuable resource for aspiring RF and Microwave engineers.

Professor Behagi's book is valuable in that it marries RF and Microwave theory with the practical examples using the Keysight's Advanced Design System (ADS) software.

ADS is one of today's most widely used Electronic Design Automation (EDA) software used by the world's leading companies to design ICs, RF Modules and boards in every smart phone, tablet, WiFi routers as well as Radar and satellite communication systems.

 RF and Microwave design techniques and ADS are also getting widespread adoption for baseband, high-speed applications. Knowing the fundamentals and practical application of RF and Microwave design with ADS will broaden your potential career opportunities.

Read these books and you'll have an advantage over others. Master all the 100 examples and additional exercises will help you to write your own ticket to success.

Joe Civello

Keysight Technologies

ADS Planning and Marketing Manager

1400 Fountaingrove Parkway

Santa Rosa, CA 95403, USA

Preface

The *100 ADS Design Examples* book is based on the author's *RF and Microwave Circuit Design* textbook written mainly for practicing engineers and university students who want to apply the RF and microwave theory to the analysis and design of RF and microwave circuits using the Keysight's Advanced Design System (ADS) software.

The author also uses other CAD techniques that may not be familiar to some engineers. This includes the frequent use of the MATLAB script capability. All the MATLAB and microwave equations needed to analyze and design the examples in the book are derived and used throughout the book. Additionally each example has an associated ADS workspace that comes in a separate package.

The examples in the book are divided into 8 chapters as follows.

1. RF and Microwave Components
2. Transmission. Line Components
3. Network Parameters and the Smith Chart
4. Resonant Circuits and Filter Design
5. Power Transfer and Impedance Matching
6. Distributed and Microstrip Impedance Matching
7. Single Stage Amplifier Design
8. Multi-Stage Amplifier Design

Learning all the example and doing the additional problems in each chapter will definitely help you to broaden your potential career opportunities.

Ali A. Behagi, Ph.D.
January 2016

Table of contents

Chapter 1: RF and Microwave Components   1

Chapter 3: Network Parameters and the Smith Chart   57

Chapter 4: Resonant Circuits and Filter Design   73

Chapter 6: Distributed and Microstrip Impedance Matching   163

Chapter 7: Single Stage Amplifier Design   199

Chapter 8: Multi-Stage Amplifier Design   **247**

Chapter 1

RF and Microwave Components

Straight Wire Inductance

A conducting wire carrying an AC current produces a changing magnetic field around the wire. According to Faraday's law the changing magnetic field induces a voltage in the wire that opposes any change in the current flow. This opposition to change is called self inductance. At high frequencies even a short piece of straight wire possesses frequency dependent resistance and inductance behaving as a circuit element.

Analysis and simulation of Straight Wire in ADS

Example 1.2-1: Calculate the reactance and inductance of a three inch length of AWG #28 copper wire in free space at 60 Hz, 500 MHz, and 1 GHz.

Solution: The procedure for the analysis and simulation of straight wire in ADS is as follows. This procedure are applied to other examples throughout the book.

- Start ADS and create a new workspace
- Name the workspace Ex1.2-1_wrk
- From the ADS Main window create a New Schematic in cell_1
- In the Schematic window, click on Insert > Template >ads-templates:S_Params
- Delete the DisplyTemplate icon
- Type in WIRE in the part selection box and place the wire component between the input and output Terminations
- Insert the input impedance, Zin1, from the Simulation-S_Param Palette to analyze the input impedance as a function of frequency.
- Double click on the S-PARAMETERS icon to set the frequency range as shown in Figure 1-1.

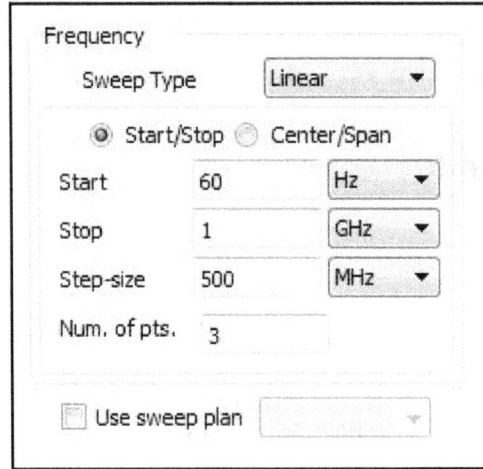

Figure 1-1 Frequency setting for simulation of wire impedance

Wire up the schematic as shown in Figure 1-2.

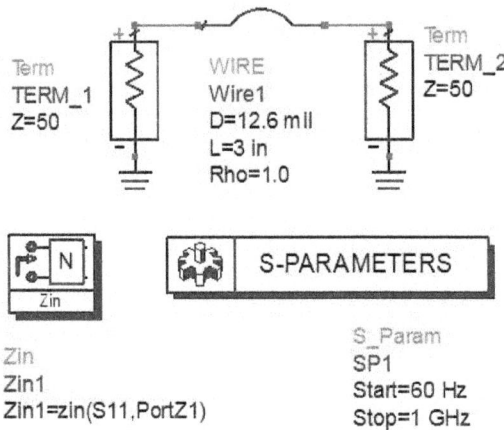

Figure 1-2 Schematic of the straight wire in ADS

In ADS, the input termination, TERM_1, represents a signal source in series with the source impedance and the output termination, TERM_2, represents the load impedance connected to ground. Set the wire diameter to 12.6 mils and the length to 3 inches. Accept the value of 1 for Rho. Rho is not the actual resistivity of the wire but rather the resistivity of the wire relative to copper. Because we are modeling a copper wire the value should be set to one.

ADS has a built-in function Zin to measure the port impedance of the circuit at each analysis frequency. Insert a built-in function Zin1 from the Simulation-S_Param palette to measure the input impedance of the wire versus frequency. When simulation is run in ADS the results are written to a Dataset. The results of a Dataset may then be sent to a graph or tabular output for visualization. More complex workspaces may have multiple Datasets. It is a good practice to specify which Dataset is used to collect data for post processing.

In this example two equations are used to post process the information in the Dataset. In the Data Display Window add equations to calculate the reactance and inductance of the straight wire, as shown here. From the "Datasets and Equations" select the Equations as the data source and then select both of the variables, reactance and inductance.

Eqn reactance=im(Zin1)

Eqn inductance=reactance/(freq*2*pi)

freq	Zin1	reactance	inductance
60.00 Hz	50.023 + j3.366E-5	3.366E-5	8.929E-8
500.0 MHz	50.538 + j280.501	280.501	8.929E-8
1.000 GHz	50.755 + j561.001	561.001	8.929E-8

Table 1-1 Input impedance, reactance and inductance

We can see that at 60 Hz the reactance is very small. At 500 MHz and 1 GHz however, the values greatly increases. This is due to the skin effect of the conductor. The skin effect is a property of conductors where, as the frequency increases, the current density concentrates on the outer surface of the conductor.

Calculation of Flat Ribbon Inductance

Flat ribbon style conductors are very common in RF and microwave engineering. Flat ribbon conductors are encountered in RF systems in the form of low inductance ground straps. Flat ribbon conductors can also be encountered in microwave integrated circuits (MIC) as gold bonding straps.

When a very low inductance is required the flat ribbon or copper strap is a good choice. The flat ribbon inductance can be calculated from the following empirical Equation.

$$L = K \ell \left[\ln\left(\frac{2\ell}{W+T}\right) + 0.223\left(\frac{W+T}{\ell}\right) + 0.5 \right] \quad nH$$

Where:

ℓ = The length of the wire
K= 2 for dimensions in cm and K=5.08 for dimensions in inches
W = the width of the conductor
T = the thickness of the conductor

Example 1.2-4: Calculate the inductance of the 3 inch Ribbon at 60 Hz, 500 MHz, and 1 GHz. Make the ribbon 100 mils wide and 2 mils thick.

Solution: The schematic of the flat ribbon is shown in Figure 1-3 along with the Zin1 and S-Parameter simulation.

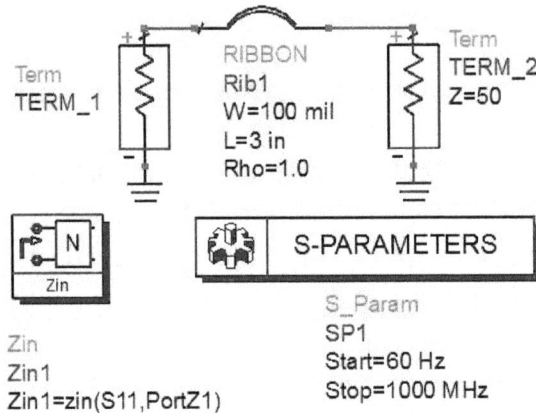

Figure 1-3 Flat ribbon schematic

Simulate the schematic to open the data display window. In the data display window add equations to calculate the reactance and inductance of the flat ribbon. Select a tabular List and add Zin1 plus reactance and inductance equations to the list, as shown here.

Eqn reactance=im(Zin1)   Eqn inductance=reactance/(freq*2*pi)

freq	Zin1	reactance	inductance
60.00 Hz	50.006 / 2.997E-5	2.616E-5	6.939E-8
500.0 MHz	220.786 / 76.883	215.026	6.844E-8
1.000 GHz	432.905 / 83.348	429.990	6.844E-8

Table 1-2. Flat ribbon tabular output Zin1, reactance, and inductance

We can see that at 60 Hz the reactance is very small. At 500 MHz and 1 GHz however, the values greatly increase. This is due to the skin effect of the conductor. The skin effect is a property of conductors where, as the frequency increases, the current density concentrates on the outer surface of the conductor.

Ideal and Physical Resistors

The resistance of a material determines the rate at which electrical energy is converted to heat. The resistivity of materials is specified in Ω-meters rather than Ω/meter. This facilitates the calculation of resistance.

Example 1.3-1A: Plot the impedance of a 50 Ω ideal resistor in ADS over a frequency range of 0 to 2 GHz.

Solution: Look at the input impedance of an ideal resistor with no parasitics at RF and microwave frequencies. Create a workspace in ADS and open a new schematic. Place a 50 Ohm resistor and measure the input impedance from 0 to 2 GHz. Complete the schematic as shown in Figure 1-4.

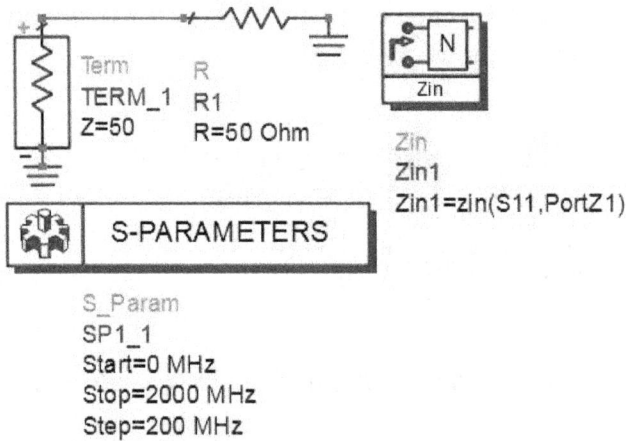

Figure 1-4 schematic of a 50 Ω ideal resistor

Simulate the schematic and display the input impedance over a frequency range of 0 to 2 GHz. The plot shows a constant resistance at all frequencies.

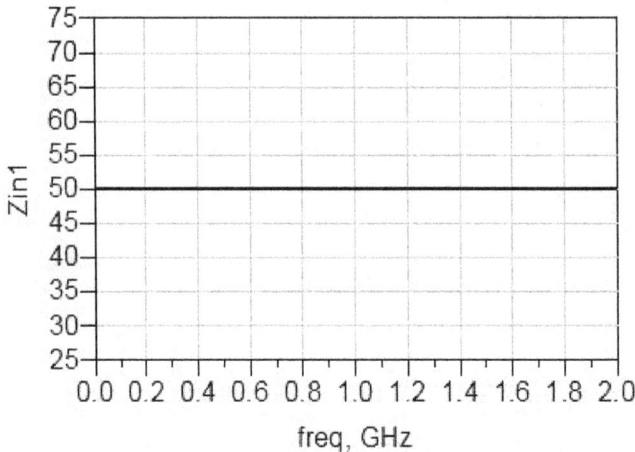

Figure 1-5 Ideal 50Ω resistor impedance versus frequency

Example 1.3-1B: Plot the impedance of a 5 Ω leaded resistor in ADS over a frequency range of 0 to 2 GHz.

Solution: At RF and microwave frequencies however, resistors also possess inductive and capacitive elements. The stray inductance and capacitance associated with a resistor are often called parasitic elements. Consider the leaded resistor model as shown in Figure 1-6. For a 1/8 watt leaded resistor it is not uncommon for each lead to have about 10 nH of inductance. The body of the resistor may exhibit 5 pF capacitance between the leads. Designing the network in ADS reveals an interesting result of the impedance versus frequency response.

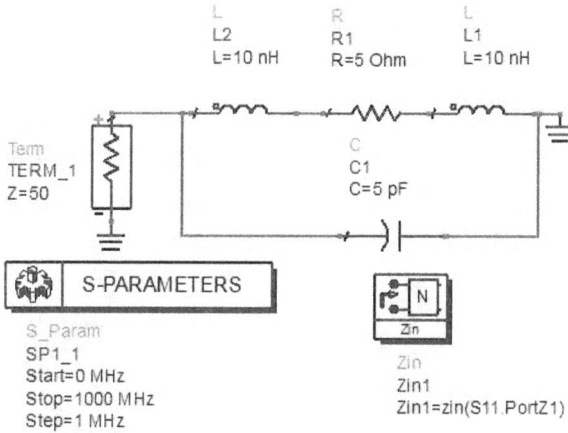

Figure 1-6 Schematic of a 5 Ω leaded resistor with parasitic elements

Simulate the schematic and plot the input impedance as shown here.

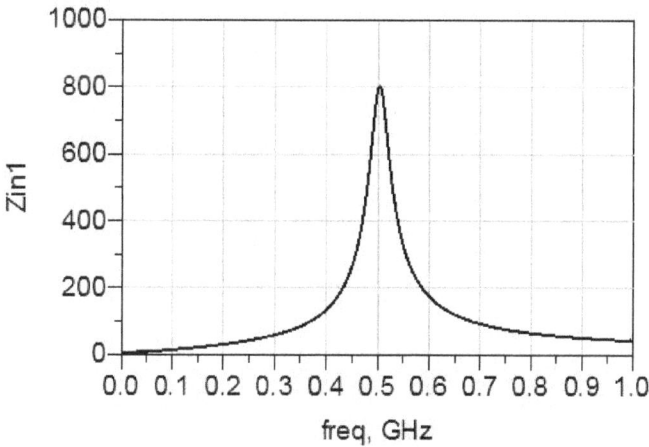

Figure 1-7 Impedance of the leaded resistor versus frequency

Notice the resonance of the resistance model at 500 MHz.

Chip Resistors

Thick film resistors are used in most contemporary electronic equipment. The thick film resistor, often called chip resistor, comes close to eliminating much of the inductance that plagues the leaded resistor. The chip resistor works well with popular surface mount assembly techniques preferred in modern electronic manufacturing.

Example 1.3-2: Plot the impedance of 1 kΩ, 0603 size chip resistor, manufactured by KOA, from 0 to 3 GHz. This model is available in the Modelithics evaluation model kit.

Solution: Create a schematic in ADS, select the Modelithics Library in the ADS Part Selector and place the KOA0603 resistor on the schematic, as shown with the arrow in Figure 1-8.

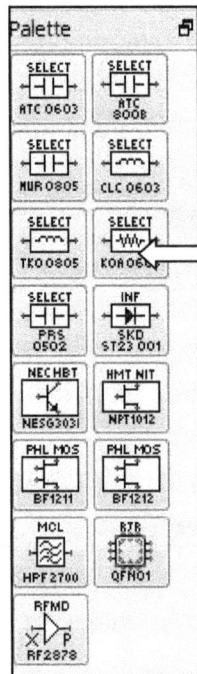

Figure 1-8 Modelithics chip component library in ADS

Complete the schematic as shown in Figure 1-9.

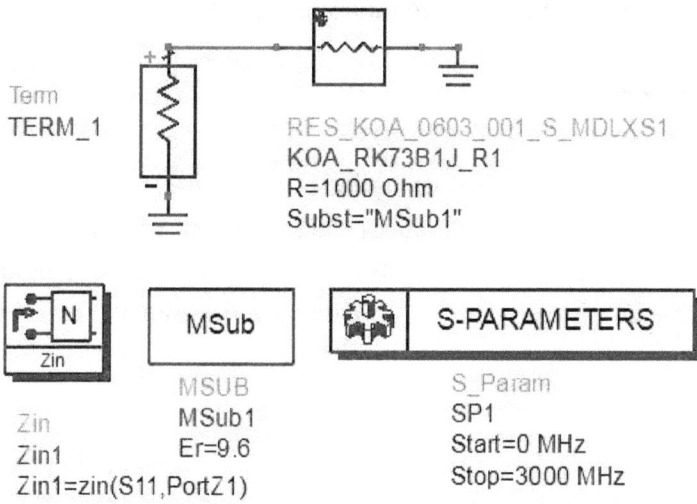

Figure 1-9 Schematic of the Modelithics chip resistor model

Simulate the schematic and display the input impedance as a function of frequency in a rectangular plot, as shown in Figure 1-10.

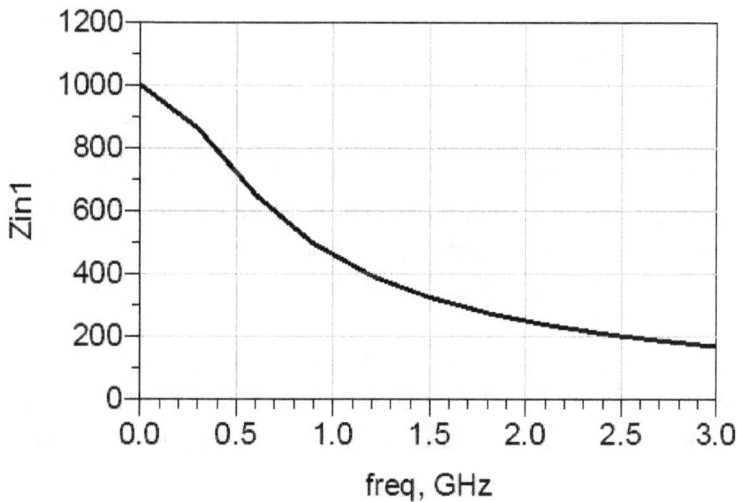

Figure 1-10 Modelithics chip resistor model impedance versus frequency

Note the roll off of the impedance with increasing frequency. This suggests that the chip resistor does have a parasitic capacitance that is in parallel with the resistor similar to the discrete model of Figure 1-6.

Air Core Inductors

Forming a wire on a removable cylinder is the basic realization of the air core inductor. When designing an air-core inductor, use the largest wire size and close spaced windings to result in the lowest series resistance and high Q. The basic empirical equation to calculate the inductance of an air core inductor is given here.

$$L = \frac{(17)N^{1.3}(D+D1)^{1.7}}{(D1+S)^{0.7}}$$

Where:

N = Number of turns of wire
D = Core form diameter in inches
$D1$ = Wire diameter in inche
L = Coil inductance in nH
S = Spacing between turns in incchess

As an interesting comparison with Example 1.2.1 calculate the amount of inductance that we can realize in that same three inches of wire if we wind it around a core to form an inductor. Choose a core form of 0.095 inches diameter as a convenient form to wrap the wire around.

First we need to calculate the approximate number of turns that we can expect to have with the three inch length of wire. We know that the circumference of a circle is related to the diameter by the following equation.

$$Circumference = \pi \ (Diameter) = \pi(0.095) = 0.2985 \text{ inches}$$

With a circumference of 0.2985 inches we can calculate the approximate number of turns that we can wrap around the 0.095 inch core with three inches of wire.

$$N = \frac{3}{0.2985} = approximately\ 10\ turns$$

From the equation for the calculation of inductance we see that the spacing between the turns has a strong effect on the value of the inductance that we can expect from the coil. When hand winding the coil, it may be difficult to maintain an exact spacing of zero inches between the turns. Therefore it is useful to solve the equation in terms of a variety of coil spacing so that we can see the effect on the inductance. We can solve the equation in MATLAB for a variety of coil spacing, as follows.

1. Enter design parameters

D=.095; D1=.0126; N=10; Spacing = [0;.002;.004;.006;.008;.010];

2. Calculate the coil length and inductance in nano Henries

Inductance_nH = (17*(N^1.3)*((D+D1)^1.7))/((D1+Spacing)^.7)

Coil_Length = (D1*N) + (Spacing.*(N-1))

Note that the coil spacing is variable and Spacing has been defined as an array variable. Placing a semicolon between the values makes the array organized in a column format. This is handy for viewing the results in tabular format. Placing a comma between the values would organize the array in a row format. As an aid in forming the coil, the overall coil length is also calculated. The coil length is simply the summation of the overall wire thickness times the number of turns and the spacing between the turns. Add the calculated inductance, spacing, and coil length to the table as shown in Table1-3.

Index	Inductance_nH	InductanceCalculator.Spacing	InductanceCalculator.Coil_Length
1	163.784	0	0.126
2	147.735	2e-3	0.144
3	135.037	4e-3	0.162
4	124.701	6e-3	0.18
5	116.097	8e-3	0.198
6	108.806	0.01	0.216

Table 1-3 Coil inductance, spacing and coil length

Modeling the Air Core Inductor in ADS

Example 1.4-2: Calculate and plot the input impedance of an air core inductor.

Solution: The air core inductor model is shown in Figure 1-11.

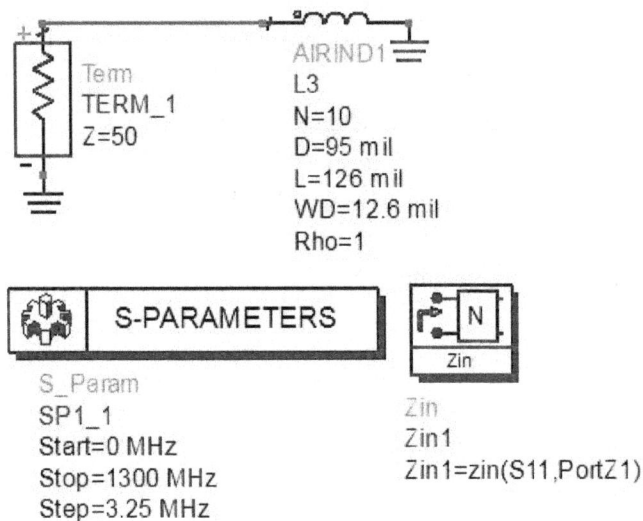

Term
TERM_1
Z=50

AIRIND1
L3
N=10
D=95 mil
L=126 mil
WD=12.6 mil
Rho=1

S-PARAMETERS

S_Param
SP1_1
Start=0 MHz
Stop=1300 MHz
Step=3.25 MHz

Zin
Zin1
Zin1=zin(S11,PortZ1)

Figure 1-11 ADS model of the air core inductor

Set up a linear frequency sweep in a range of 0 to 1.3 GHz. Plot the impedance of the inductor across the frequency range as shown in Figure 1-12. Note the interesting spike, or increase in impedance that occurs around 1.170 GHz. This is the parallel, self resonant frequency, of the inductor.

m1
freq=1.170GHz
Zin1=3.643E5 / -7.458
Peak

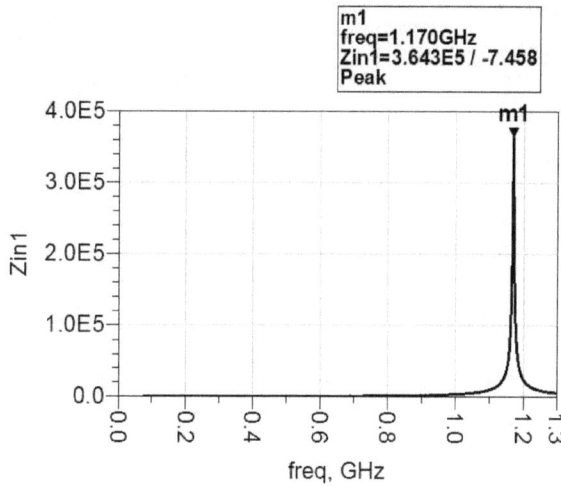

Figure 1-12 Impedance of the air core inductor as a function of frequency

Note that the inductor is not an ideal component or a pure inductance but rather a network that includes parasitic capacitance and resistance.

Example 1.4-3 Create a simple RLC network that gives an equivalent impedance response similar to Figure 1-12. One such circuit is shown in Figure 1-13.

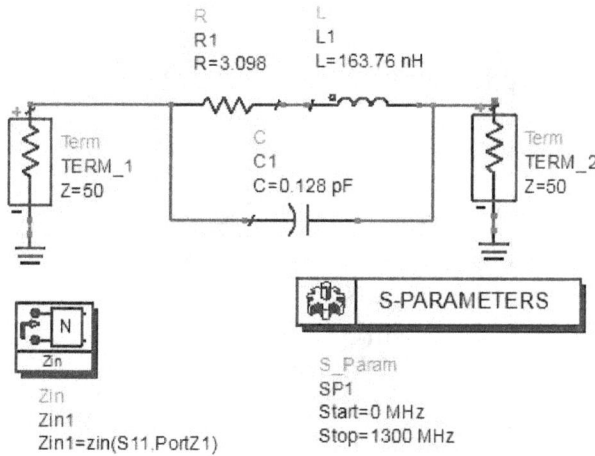

Figure 1-13 Equivalent network of the air core inductor

The simulated response is given in Figure 1-14.

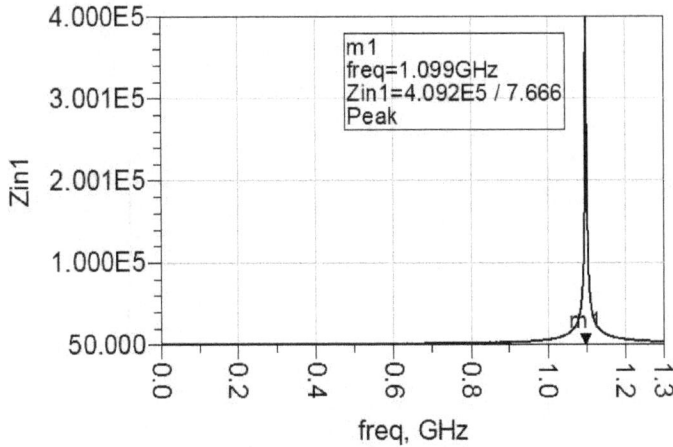

Figure 1-14 Input impedance of equivalent air core inductor

1.4.3 Inductor Q Factor

Example 1.4-4: Calculate the Q factor of the air core inductor shown in Figure 1-11.

Solution: The plot of inductor Q factor (Q=|X/R|) for the air core inductor model in Figure 1-11 is shown in Figure 1-15.

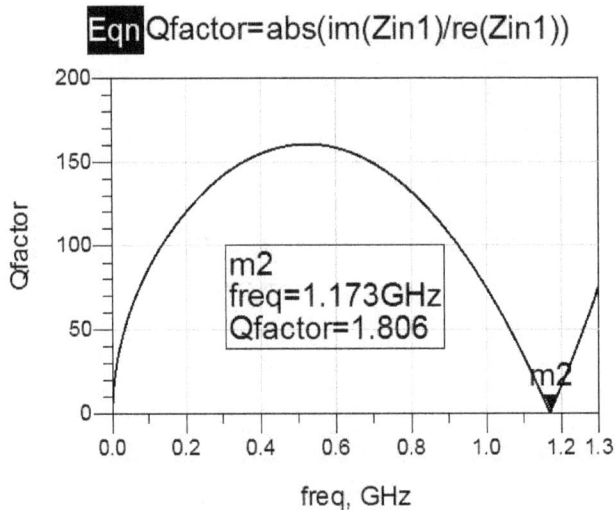

Figure 1-15 Air core inductor model Q factor versus frequency

It is interesting to note that the Q factor peaks at a frequency well below the self-resonant frequency of the inductor. The actual frequency at which the Q factor peaks will vary among inductor designs but is usually ranges from 2 to 5 times less than the SRF. Close winding spacing results in inter-winding capacitance, which lowers the self-resonant frequency of the inductor. Thus there is a tradeoff between maximum Q factor and high self-resonant frequency. It is also noteworthy that the Q factor goes to zero at the self-resonant frequency (see marker m2 on Figure 1-15).

Chip Inductor Simulation in ADS

The ADS library has a collection of S parameter files for the Coilcraft chip inductors.

Example 1.4-5: Calculate and plot the inductance and Q factor of the Coil Craft 180 nH series 0603CS chip inductor from 1 MHz to 1040 MHz.

Solution: From the ADS schematic window select MDLX_SELECTv9.0 palette and insert CLC0603 component. Change the inductor value to L=180 nH as shown in Figure 1-16.

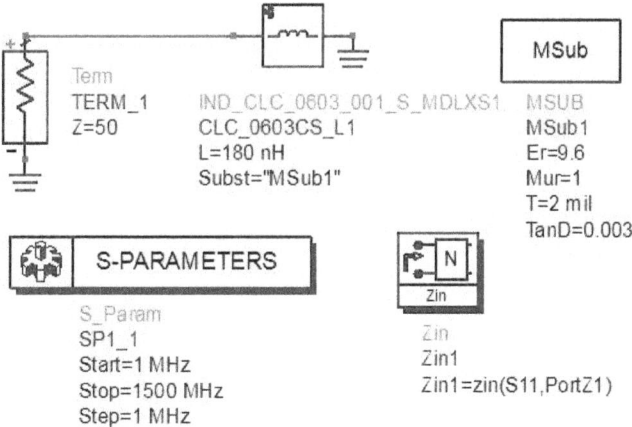

Figure 1-16 Coil Craft chip inductor and the circuit schematic

Simulate the schematic and display the inductance in a rectangular graph as shown in Figure 1-17. Place a marker at 25 MHz and at the self resonance frequency.

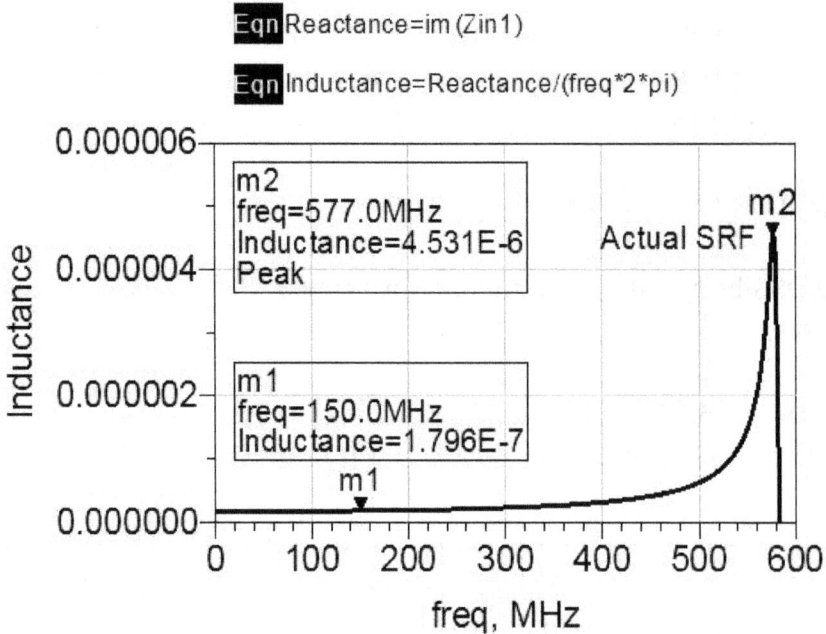

Eqn Reactance=im (Zin1)

Eqn Inductance=Reactance/(freq*2*pi)

m2
freq=577.0MHz
Inductance=4.531E-6
Peak

Actual SRF

m2

m1
freq=150.0MHz
Inductance=1.796E-7

m1

(y-axis: Inductance — 0.000006, 0.000004, 0.000002, 0.000000)
(x-axis: freq, MHz — 0, 100, 200, 300, 400, 500, 600)

Figure 1-17 Plot of the 180 nH chip inductor inductance

From the markers on the plot we see that the simulation of the S parameter file measures 179.6 nH at 150 MHz and the SRF= 577 MHz. The manufacturer specifies a minimum Q factor of 45 at 100 MHz.

Eqn Qfactor=abs(Reactance/Resistance)

m1
indep(m1)=1.920E8
plot_vs(Qfactor, freq)=41.262
Peak

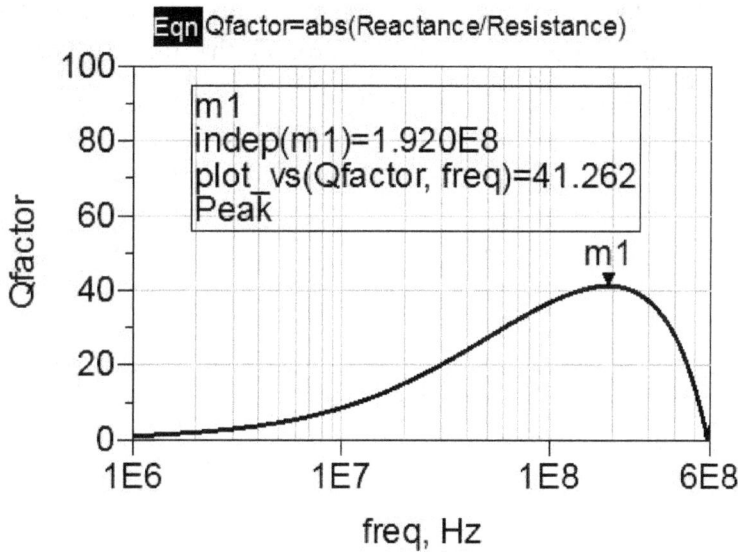

Figure 1-18 180 nH chip inductor Q factor versus frequency

The plot of the Q factor derived from the S parameter file shows that the maximum Q factor is 41.262 near 192 MHz. Manufacturers will often plot the Q vs frequency on a logarithmic scale. It is very easy to change the x-axis to a logarithmic scale on the rectangular graph properties window in ADS. This allows us to have a visual comparison with the manufacturer's catalog plot.

Magnetic Core Inductors

Example 1.4-6: Design a 550 nH inductor using the Carbonyl W core of size T30. Determine the number of turns and model the inductor in ADS.

Solution: From the manufacturer's data sheet the A_L value is 2.5 for a T30-10 toroidal core. Rearranging the Equation ($L=N^2 A_L$ nH) to solve for the number of turns we find that 14.8 turns are required.

$$N=\sqrt{\frac{L}{A_L}} \quad \sqrt{\frac{550\ nH}{2.5}} = 14.8$$

To reduce the winding loss we want to use the largest diameter of wire that will result in a single layer winding around the toroid. The following Equation will give us the wire diameter.

$$d = \frac{\pi \; ID}{N + \pi}$$

Where:

> d = Diameter of the wire in inches
> ID = Inner diameter of the core in inches
> N = Number of turns

Therefore,

$$d = \frac{\pi \; (0.151)}{14.8 + \pi} = \frac{0.4744}{19.942} = 0.0238 \quad inches$$

Normally AWG#24 is chosen because this is a more readily available standard wire size. The toroidal inductor model in ADS requires a few more pieces of information. The ADS model requires that we enter the total winding resistance, core Q factor, and the frequency for the Q factor, F_q. As an approximation, set F_q to about six times the frequency of operation. In this case set F_q to (6)(25) MHz = 150 MHz. Then tune the value of Q to get the best curve fit to the manufacturer's Q curve. We know that we have 14.8 turns on the toroid but we need to calculate the length of wire that these turns represent. The approximate wire length around one turn of the toroid is calculated from the following equation.

$$Length = \left[(2) \, Height + (OD - ID) \right] (\#turns)$$

Using the dimensions for the T30 toroid we can calculate the total length of the wire as 6.10 inches.

$$\left[(2)(0.128) + (0.307 - 0.151) \right] \cdot 14.8 = 6.10 \quad inches$$

Then use the techniques covered earlier to calculate the resistance of the 6.10 inches of wire taking into account the skin effect. To get a better estimate of the actual inductor Q, use the F_q frequency rather than the operating frequency for the skin effect calculation. The resistance at 150 MHz for the AWG#24 wire is calculated as 0.30 Ω. Figure 1-19 shows the schematic of the toroidal inductor and the simulated Q.

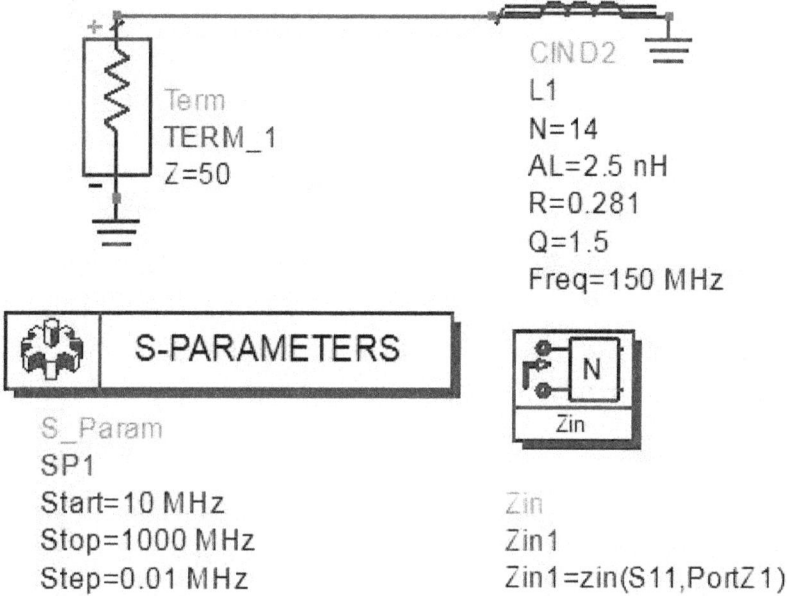

CIND2

Term
TERM_1
Z=50

L1
N=14
AL=2.5 nH
R=0.281
Q=1.5
Freq=150 MHz

S-PARAMETERS

Zin

S_Param
SP1
Start=10 MHz
Stop=1000 MHz
Step=0.01 MHz

Zin
Zin1
Zin1=zin(S11,PortZ1)

Figure 1-19 ADS toroidal inductor model

The simulated Q factor is shown in Figure 1-20.

Eqn X=im(Zin1) Eqn R=re(Zin1) Eqn Q=abs(X/R)

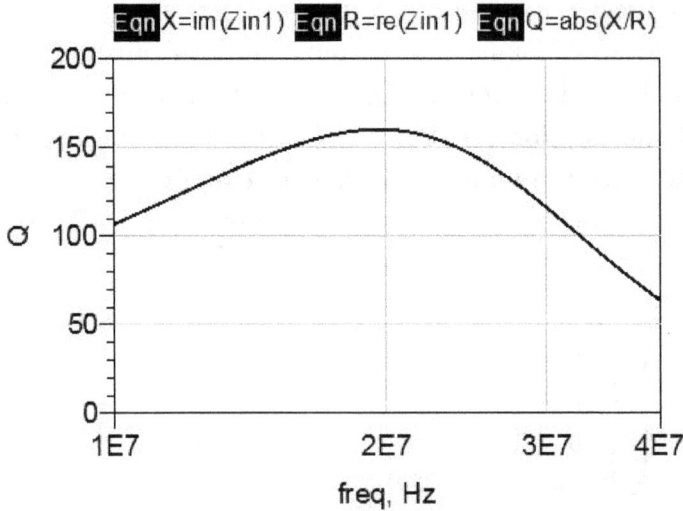

Figure 1-20 ADS toroidal inductor simulated Q factor

The Q parameter of the model has been tuned to give a reasonably good fit with the manufacturer's (T30) curve. Figure 1-21 shows that the simulated self resonant frequency of the inductor model is near 155 MHz.

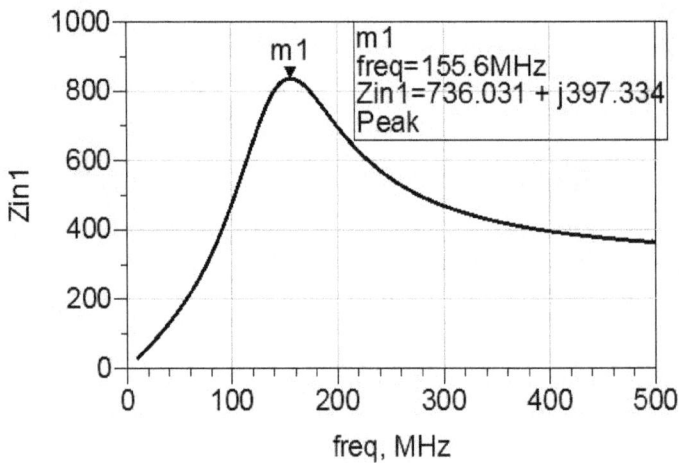

m1
freq=155.6MHz
Zin1=736.031 + j397.334
Peak

Figure 1-21 Impedance of toroidal inductor model

Figure 1-22 shows the inductance that was calculated from the impedance. The inductance is exactly 550 nH at 10 MHz and begins to increase slightly to 563 nH at the design frequency of 25 MHz. Given the myriad of variables associated with the toroidal inductor, this simple model gives a good first order model of the actual inductor and will enable accurate simulation of filter or resonator circuits.

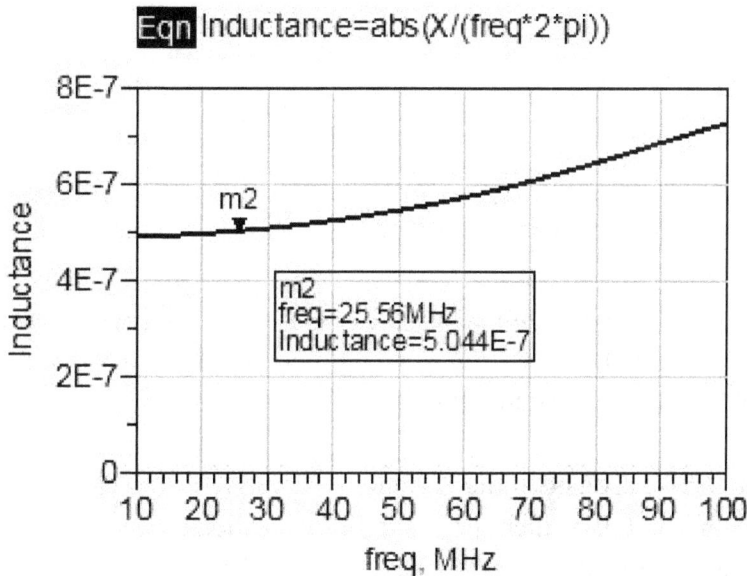

Figure 1-22 Inductance value of the toroidal inductor model

Single Layer Capacitor

The single layer capacitor is one of the simplest and most versatile of the surface mount capacitors. It is formed with two plates that are separated by a single dielectric layer as shown in Figure 1-23. Most of the electric field (E) is contained within the dielectric however there is a fraction of the E field that exists outside of the plates. This is known as the fringing field.

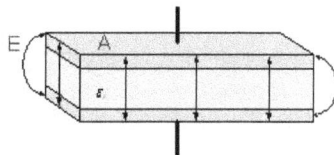

Figure 1-23 Single layer parallel plate capacitor

The capacitance formed by a dielectric material between two parallel plate conductors is given:

$$C = (N-1)\left(\frac{KA\,\varepsilon_{\mathrm{r}}}{t}\right)(FF) \quad pF$$

Where,

A = plate area

ε_r = relative dielectric constant

t = separation

K = unit conversion factor; 0.885 for cm and 0.225 for inches

FF = fringing factor; 1.2 when mounted on microstrip

N = number of parallel plates.

Example 1.5-1: Consider the design of a single layer capacitor from a dielectric that is 0.010 inches thick and has a dielectric constant of three. Each plate is cut to 0.040 inches square. Calculate the capacitor value and its Q factor.

Solution: When the capacitor is mounted with at least one plate on a large printed circuit board track, a value of 1.2 is typically used in calculation for the fringing factor.

$$C = (2-1)\left(\frac{(.225)(0.04 \cdot 0.04)(3)}{0.010}\right)(1.2) = 0.13\,pF$$

The single layer capacitor can be modeled in ADS using the Thin Film Capacitor, as shown in Figure 1-24.

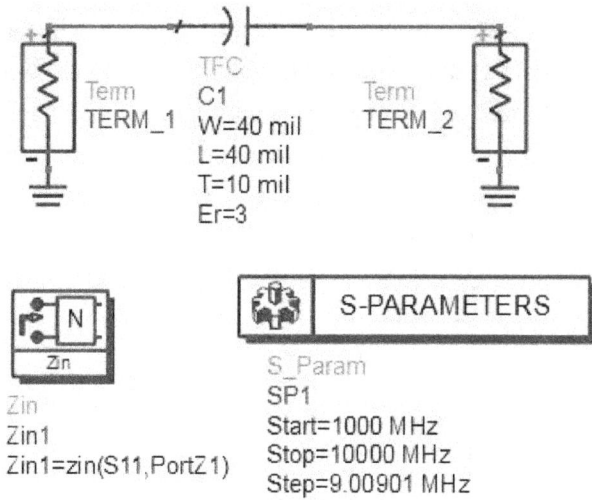

Figure 1-24 Single layer parallel plate capacitor model in ADS

The capacitor Q factor versus frequency is given by simulating the schematic and plotting the Qfactor = abs(X/R).

Eqn Reactance=im(Zin1)   Eqn Resistance=re(Zin1)

Eqn Qfactor=abs(Reactance)/Resistance

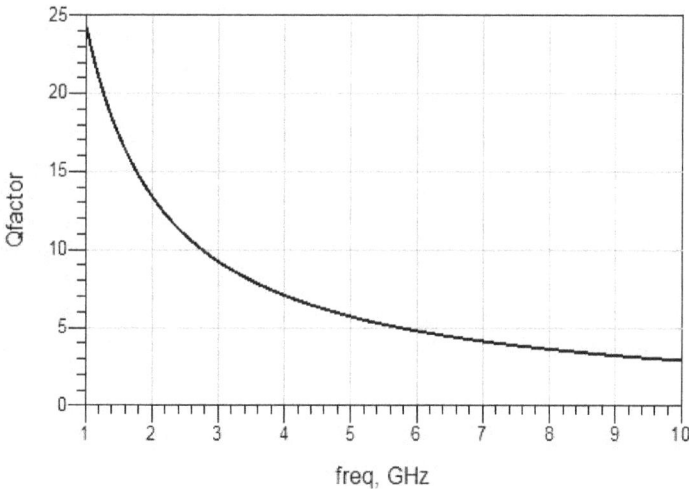

Figure 1-25 Single-layer Capacitor Q factor

Capacitor Physical Model and Q Factor

RF losses in the dielectric material of a capacitor are characterized by the dissipation factor. The dissipation factor is also referred to as the loss tangent and is the ratio of energy dissipated to the energy stored over a period of time. It is essentially the capacitor's efficiency rating. The dissipation factor and other ohmic losses lead to a parameter known as the Equivalent Series Resistance, ESR. The dissipation factor is the reciprocal of the Q factor. Just as we have seen with resistors and inductors, the physical model of a capacitor is a network of R, L, and C components.

Example 1.5-2A: Calculate the Q factor versus frequency for the physical model of an 8.2 pF multilayer chip capacitor shown in Fig. 1-26.

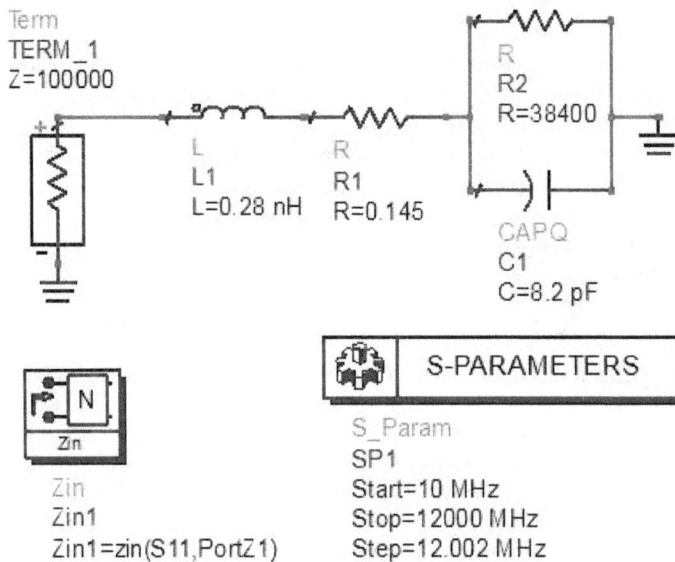

Figure 1-26 Physical model of the 8.2 pF chip capacitor

Solution: The impedance of the physical model is shown in Figure 1-27.

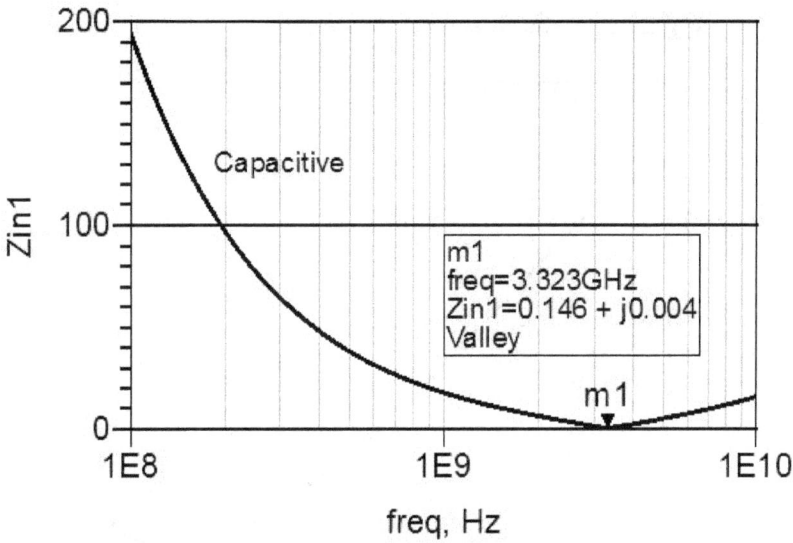

Figure 1-27 Physical 8.2 pF chip capacitor impedance

The capacitor has a series inductance and resistance component along with a resistance in parallel with the capacitance. The parallel resistor sets the losses in the dielectric material. The series resistance and inductance represent any residual lead inductance and ohmic resistances.

The inductance, capacitance, and Q factor can be calculated from the impedance using the following equations.

Reactance = im(Zin1)

Resistance = re(Zin1)

Inductance = abs((Reactance)/(freq*2*pi))

Qfactor = abs(Inductance/Resistance)

Capacitance = 1/(abs(Reactance)*freq*2*pi)

Therefore, the Q factor plot is shown in Figure 1-25.

Eqn Qfactor=abs(Reactance/Resistance)

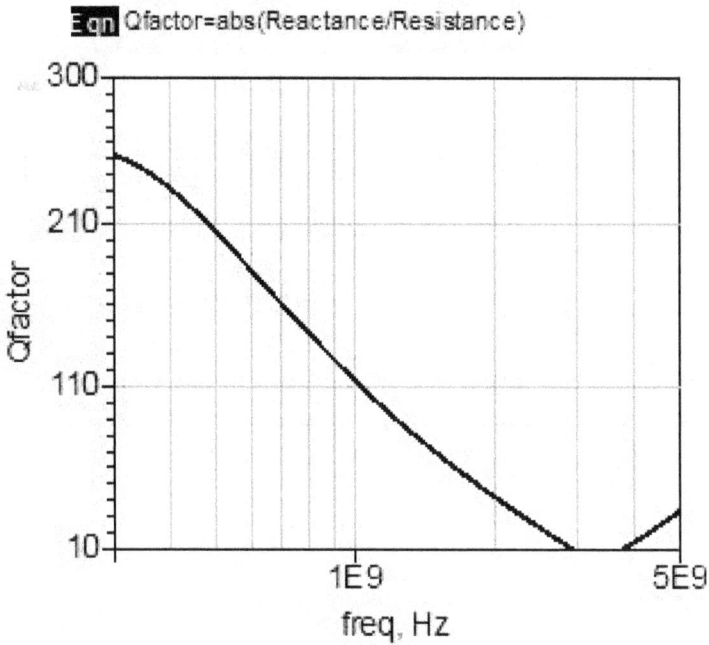

Figure 1-28: Q factor of the Physical Model

Example 1.5-2B: Calculate the Q factor versus frequency for the modified physical model of an 8.2 pF multilayer chip capacitor shown in Fig. 1-23.

Solution: The modified model is shown in Figure 1-29.

Term   L   R   CAPQ
TERM_1   L1   R1   C2
Z=10000   L=0.28 nH   R=0.145 Ohm   C=8.2 pF
                                 Q=3000
                                 F=1 MHz

Zin
Zin1
Zin1=zin(S11,PortZ1)

S-PARAMETERS

S_Param
SP1
Start=100 MHz
Stop=100 GHz

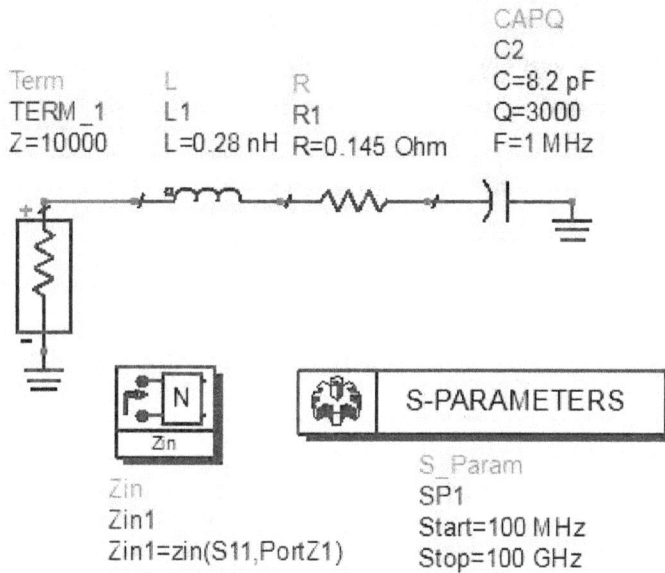

Figure 1-29 Modified physical model of 8.2 pF chip capacitor

The capacitor Q factor of the modified model versus frequency is plotted in Figure 1-30.

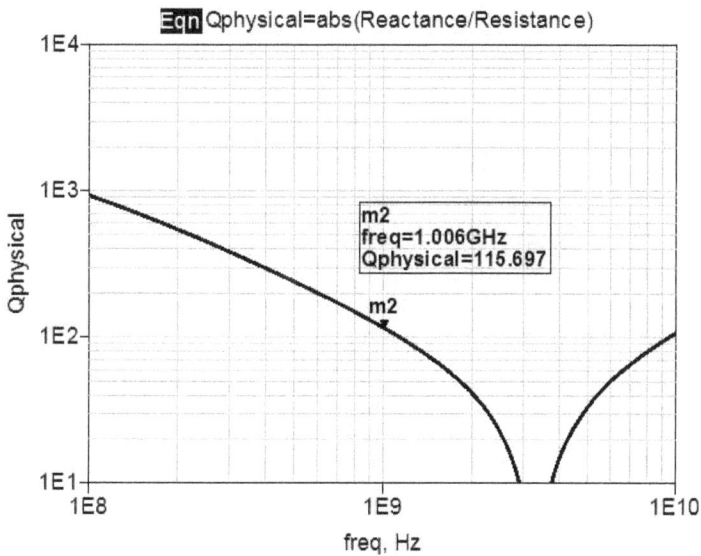

Eqn Qphysical=abs(Reactance/Resistance)

m2
freq=1.006GHz
Qphysical=115.697

m2

Qphysical

freq, Hz

Figure 1-30 Calculated Q factor of 8.2 pF chip capacitor

Figure 1-30 shows the large dependence of the capacitor Q on frequency. The Q factor goes to zero at the self resonant frequency. Above the self resonant frequency the Q factor is undefined.

Calculate the effective capacitance from the total reactance of the model using the following equation.

$$C = \frac{1}{2\pi F X_T}$$

The plot of Figure 1-31 reveals some interesting characteristics about the chip capacitor. From 100 MHz to 300 MHz the capacitance value is fairly constant. As the frequency increases we see that the capacitance actually increases. The parasitic inductive reactance of the capacitor package actually makes the effective capacitance greater than its nominal value.

Eqn Capacitance=1/(abs(Reactance)*2*pi*freq)

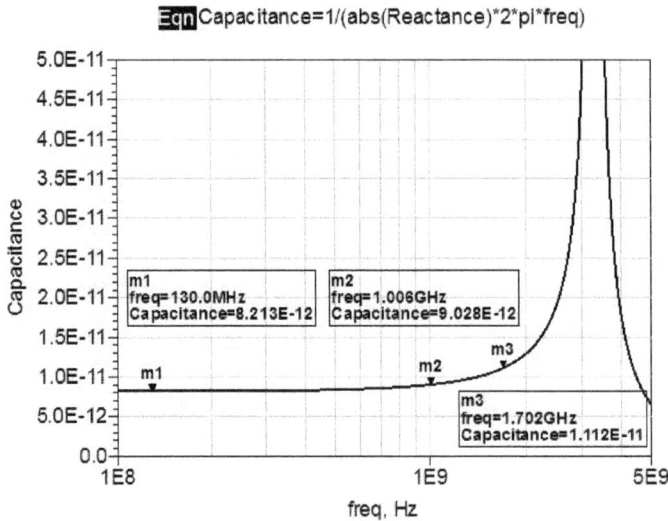

Figure 1-31 Effective capacitance of the 8.2 pF chip capacitor

References and Further Reading

[1]　　Ali A. Behagi, *RF and Microwave Circuit Design,* A Design Approach Using (**ADS**). Techno Search, Ladera Ranch, CA 92694. August 2015.

[2]　　Paul Lorrain, Dale P. Corson, and Francois Lorrain, *Electromagnetic Fields and Waves*, W.H. Freeman and Company, New York, 1988

[3]　　*Design Guide, Microwave Components* Inc., P.O. Box 4132, South Chelmsford, MA 01824

[4]　　*Iron Powder Cores for High Q Inductors*, Micrometals, Inc.

[5]　　Keysight Technologies, Manuals for Advanced Design System, *ADS 2016.01 Documentation Set*, Keysight EEsof EDA Division, Santa Rosa, California, www.keysight.com

[6]　　*Capacitors for RF Applications,* Dielectric Laboratories, Inc.,2777 Rt.20 East, Cazenovia, NY. 13035

[7]　　R. Ludwig and P. Bretchko, *RF Circuit Design -Theory and Applications*, Prentice Hall, New Jersey, 2000

[8]　　William Sinnema and Robert McPherson, *Electronic Communication,* Prentice Hall Canada, 1991

[9]　　M.F. "Doug" DeMaw, *Ferromagnetic Core Design & Application Handbook*, MFJ Publishing Co., Inc. Starkville, MS. 39759, 1996

[10]　　*The RF Capacitor Handbook,* American Technical Ceramics, One Norden Lane, Huntington, New York 11746

Problems

1-1.   Calculate the wavelength of an electromagnetic wave operating at a frequency of 428MHz.

1-2.   Calculate the inductance of a 5 inch length of AWG #30 straight copper wire.

1-3.   Calculate the reactance of the wire from Problem 2 at 10 Hz, 10 MHz, and 10 GHz. Create a Linear analysis in ADS and display the wire impedance vs. frequency.

1-4.   Calculate the resistance of a 12 inch length of AWG #24 copper wire at DC and at 25MHz

1-5.   Find the skin depth and the resistance of a 2 meter length of copper coaxial line at 2 GHz. The inner conductor radius is 1 mm and the outer conductor is 4 mm.

1-6.   Calculate the inductance of a 5 inch length of copper flat ribbon conductor. The dimensions of the ribbon are 0.100 inches in width and 0.002 inches thick.

1-7.   Model a chip resistor (size 0603) with a resistance of 50Ω in ADS. Consider an application in which 50Ω impedance must be maintained with +10%. Create a linear analysis and determine the maximum usable frequency of the chip resistor.

1-8.   Design an air core inductor with an inductance value of 84 nH. Use a copper wire of 0.050 inch diameter wound on a core diameter or 0.100 inch. Determine the number of turns required assuming a tight spaced winding.

1-9.   Using the inductor from Problem 1-8 examine the change in the coil inductance as the turn spacing is increased from zero to 0.10 inch in 0.002 inch increments.

1-10. Using the inductor from Problem 8, determine the self resonant frequency of the inductor and comment on the maximum frequency in which the inductor may be used in a tuned circuit application.

1-11. Using the inductor from Problem 1-8, determine the maximum Q factor of the inductor and the frequency at which the maximum Q factor is obtained.

1-12. In ADS, select a chip inductor from the CoilCraft library with an inductance value of 80nH. Determine the maximum Q factor of the inductor and comment on the maximum usable frequency of the inductor in a filter application.

1-13. Design a 1 mH toroidal inductor on a Carbonyl W core size T30. Determine the maximum wire size that could be used to realize a single layer winding.

1-14. Using the inductor from Problem 1-13, model the inductor in ADS and determine the approximate self resonant frequency. Comment on the maximum usable frequency of the inductor in the front end of a radio receiver.

1-15. A 0.05pF capacitor is required to couple a transistor to the resonator of a microwave oscillator. Design a single layer capacitor using a 0.020 inch thick dielectric with e_r=2.2. Determine the dimensions of the capacitor assuming square footprint is desired.

1-16. For the single layer capacitor of Problem 15, determine the dimensions of the capacitor with a dielectric constant e_r=10.2.

1-17. In ADS, model a 47pF chip capacitor using the ATC 0603 model from the Modelithics library. Determine the Q factor of the capacitor at a frequency of 1000MHz.

1-18. Determine the self resonant frequency of the capacitor in Problem 17 and comment on the maximum frequency that this capacitor could be used in a tuned circuit. At what frequency does the capacitor have the lowest amount of energy loss?

Chapter 2

Transmission Lines

Introduction

Transmission lines play an important role in designing RF and microwave networks. In chapter 1 we have seen that, at high frequencies where the wavelength of the signal is smaller than the dimension of the components, even a small piece of wire acts as an inductor and affects the performance of the network.

Example 2.4-1: For the series RLC elements in Figure 2-1 measure the reflection coefficients and VSWR from 100 to 1000 MHz in 100 MHz steps.

Solution: The procedure for the analysis of Example 2.4-1 in ADS is as follows. Similar procedure will be repeated throughout the book.

- Start ADS and create a new workspace
- Name the workspace Ex2.4-1_wrk
- From the Main window create a new schematic in cell_1
- In the schematic window, click Insert and select S_Params Template
- Delete the DisplyTemplate icon
- Type in SRLC in the part selection box and place the series RLC component between the input and output Terminations
- Set R=55 Ω, L=15 nH, and C=20 pF
- Type in VSWR in the part selection box and place it on the schematic to measure the voltage standing wave ratio as a function of frequency.
- Set the frequency range from 100 to 1000 MHz in 100 MHz step and wire up the schematic as shown in Figure 2-1

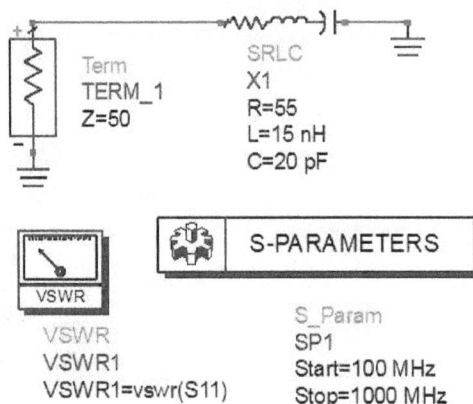

Figure 2-1 Schematic of the series RLC resonator

Simulate the schematic to measure the VSWR1 at the input port and the reflection coefficient, S(1,1), in dB and S(1,1) in Mag/Degrees formats, as shown in Table 2-1.

freq	VSWR1	dB(S(1,1))	S(1,1)
100.0 MHz	3.514	-5.084	0.557 / -52.176
200.0 MHz	1.503	-13.933	0.201 / -65.292
300.0 MHz	1.106	-25.945	0.050 / 18.321
400.0 MHz	1.420	-15.207	0.174 / 64.690
500.0 MHz	1.811	-10.796	0.289 / 64.345
600.0 MHz	2.245	-8.321	0.384 / 61.007
700.0 MHz	2.727	-6.682	0.463 / 57.292
800.0 MHz	3.260	-5.506	0.531 / 53.694
900.0 MHz	3.849	-4.620	0.588 / 50.344
1.000 GHz	4.494	-3.931	0.636 / 47.270

Table 2-1 Tabular output of VSWR, and reflection coefficient

Return Loss, VSWR, and Reflection Coefficient Conversion

Return Loss, VSWR, and Reflection Coefficient are all different ways of characterizing the wave reflection. These definitions are often used interchangeably in practice.

Example 2.4-2: Generate a table showing the return loss, the reflection coefficient, and the percentage of reflected power as a function of VSWR.

Solution: In ADS create a schematic with a resistor as shown in Figure 2-2. Make the resistance value a tunable variable. Set the Linear Analysis at a fixed frequency of 100 MHz. Then add a Parameter Sweep to the ADS Workspace to sweep the value of the resistor.

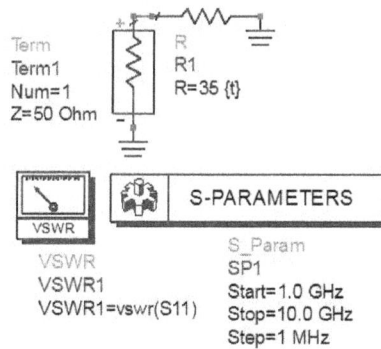

Figure 2-2 Schematic and parameter sweep for VSWR table

Edit the Parameter Sweep and select the current Linear Analysis. Then select the resistance of the Resistor element on the Parameter to Sweep drop down list. Under the Type of Sweep choose the List option and enter the discrete resistance values as shown in Figure 2-3. When the circuit is swept the 30 resistance values will create a unique VSWR, Return Loss, and Reflection Coefficient. Because ADS has no built-in function for the mismatch loss we will need to create an equation to calculate this parameter. Following are the equations for the calculation of mismatch loss.

ReflCoef = (VSWR - 1)/(VSWR + 1)
Mismatch = 1 – ((ReflCoef)^2)
Powerloss = (1 – Mismatch)*100

We will present the mismatch loss as a percentage of the available power that is reflected by the load. Add the equation block variable, power loss, to the output table. The results of the ADS Table can be saved to a comma delimited text file. This text file can then be read into Excel for a more attractive formatted table as shown in Table 2-2. As we can see from the table if we can keep the VSWR less than 1.25:1 we will have less than 1% power loss due to reflective impedance mismatch.

VSWR	Return Loss (dB)	$|\Gamma|$	% Reflected Power	VSWR	Return Loss (dB)	$|\Gamma|$	% Reflected Power
1.01	-46.06	0.00	0.00%	2.40	-7.71	0.41	16.96%
1.02	-40.09	0.01	0.01%	2.50	-7.36	0.43	18.37%
1.10	-26.44	0.05	0.23%	3.00	-6.02	0.50	25.00%
1.20	-20.83	0.09	0.83%	3.50	-5.11	0.56	30.86%
1.30	-17.69	0.13	1.70%	4.00	-4.44	0.60	36.00%
1.40	-15.56	0.17	2.78%	4.50	-3.93	0.64	40.50%
1.50	-13.98	0.20	4.00%	5.00	-3.52	0.67	44.44%
1.60	-12.74	0.23	5.33%	6.00	-2.92	0.71	51.02%
1.70	-11.73	0.26	6.72%	7.00	-2.50	0.75	56.25%
1.80	-10.88	0.29	8.16%	8.00	-2.18	0.78	60.49%
1.90	-10.16	0.31	9.63%	9.00	-1.94	0.80	64.00%
2.00	-9.54	0.33	11.11%	10.00	-1.74	0.82	66.94%
2.10	-9.00	0.36	12.59%	20.00	-0.87	0.91	81.86%
2.20	-8.52	0.38	14.06%	200.00	-0.09	0.99	98.02%
2.30	-8.09	0.39	15.52%	2000.00	-0.01	1.00	99.80%

Table 2-2 Relationship among return loss, VSWR, and reflection coefficient

Waveguide Transmission Lines in ADS

Because of the complex EM fields that can propagate in a waveguide, modern computer aided design techniques are best handled by three dimensional EM solvers. ADS have two models of the waveguides that are useful to engineers. The first model is a straight section of the waveguide in which the $TE_{1,0}$ mode is utilized. The second model is a waveguide to TEM transition which is similar to an adapter. The input and output ports used in ADS can be thought of as coaxial ports supporting TEM propagation. Therefore we cannot attach a section of waveguide directly to a port because an impedance mismatch will exist.

The cross sectional dimension of commonly used rectangular waveguide is given a WR designator. Table 2-3 shows a listing of some of the more popular waveguides by WR designator.

Frequency Band, GHz	U.S. (EIA) Designator	British WG Designator	Cut Off Freq. in GHz $TE_{1,0}$	a dimension inches	b dimension inches
1.12 - 1.70	WR 650	WG 6	0.908	6.500	3.250
1.45 - 2.20	WR 510	WG 7	1.158	5.100	2.550
1.70 - 2.60	WR 430	WG 8	1.375	4.300	2.150
2.20 - 3.30	WR 340	WG 9A	1.737	3.400	1.700
2.60 - 3.95	WR 284	WG 10	2.080	2.840	1.340
3.30 - 4.90	WR 229	WG 11A	2.579	2.290	1.145
3.95 - 5.85	WR 187	WG 12	3.155	1.872	0.872
4.90 - 7.05	WR 159	WG 13	3.714	1.590	0.795
5.85 - 8.20	WR 137	WG 14	4.285	1.372	0.622
7.05 - 10.00	WR 112	WG 15	5.260	1.122	0.497
8.2 - 12.4	WR 90	WG 16	6.560	0.900	0.400
9.84 - 15.0	WR 75	WG 17	7.873	0.750	0.375
11.9 - 18.0	WR 62	WG 18	9.490	0.622	0.311
14.5 - 22.0	WR 51	WG 19	11.578	0.510	0.255
17.6 - 26.7	WR 42	WG 20	14.080	0.420	0.170
21.7 - 33.0	WR 34	WG 21	17.368	0.340	0.170
26.4 - 40.0	WR 28	WG 22	21.100	0.280	0.140
32.9 - 50.1	WR 22	WG 23	26.350	0.224	0.112
39.2 - 59.6	WR 19	WG 24	31.410	0.188	0.094
49.8 - 75.8	WR 15	WG 25	39.900	0.148	0.074
60.5 - 91.9	WR 12	WG 26	48.400	0.122	0.061
73.8 - 112.0	WR 10	WG 27	59.050	0.100	0.050

Table 2-3 Standard rectangular waveguide characteristics

To interface waveguide with coaxial, microstrip, or stripline transmission lines a special transformer, known as an adapter, must be used. Waveguide to coax adapters come in many forms. Some adapters couple energy to the *E* field while others couple to the *H* field.

Example 2.9-1 Consider the model of a one inch and a three inch length of the waveguide as used in an X Band satellite transmit system. Display the insertion loss of the waveguides from 4 to 8 GHz.

Solution: Make the length of the waveguide tunable. From Table 2-3 we can get the width (a) and height (b) dimensions to enter into the waveguide model. Set the source and load resistors equal to 377 Ω (simulating the waveguide to TEM adapter) representing the impedance of free space. Sweep the insertion loss (S21) from 4 GHz to 8 GHz as shown in Figures 2-4 and 2-5. Note that the use of the waveguide models does require a substrate definition. In this case however the waveguide models only use the dielectric constant Er, and Rho, the resistivity of the metal walls. Typically enter values of one for both the air dielectric and the resistivity normalized to copper. Figure 2-3 shows the schematic of the waveguide. The insertion loss of the waveguide in its pass band at 8 GHz is extremely low. This is one of the advantages of using waveguide transmission lines as they are practically the lowest loss microwave transmission line available. Also note that the insertion loss increases as we move below the cutoff frequency. A marker is placed at the cutoff frequency of 5.539 GHz. Increase the length of the waveguide to 3 inches. Note the dramatic increase in the rejection below the cutoff frequency. The insertion loss in band is still quite low. We can see that using the waveguide below the cutoff frequency is an effective method of achieving a very good microwave high pass filter.

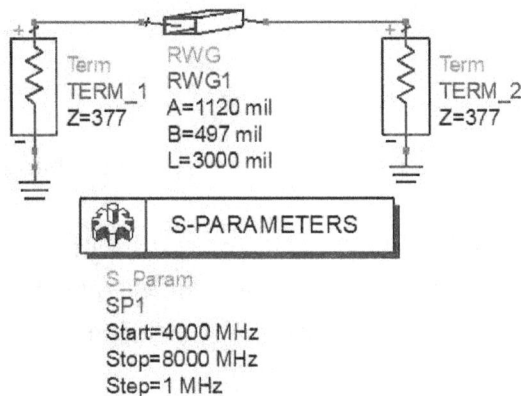

Figure 2-3 One inch length of RWG waveguide

The Insertion loss of one inch RWG waveguide is shown in Figure 2-4.

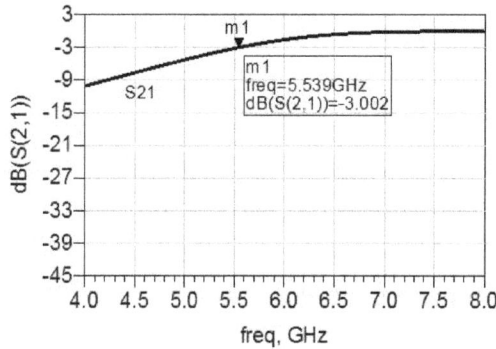

Figure 2-4 Insertion loss of one inch length of RWG waveguide

The Insertion loss of 3 inch RWG waveguide is shown in Figure 2-5.

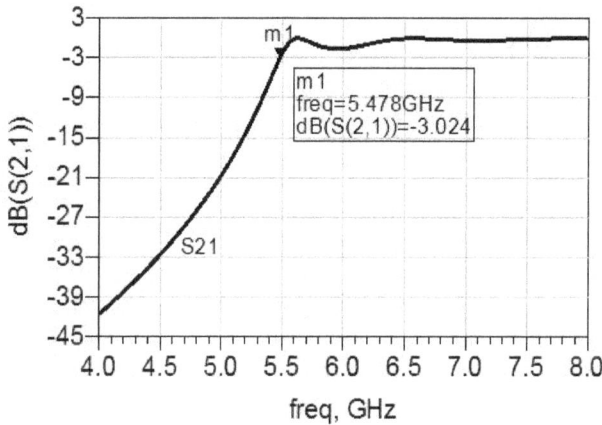

Figure 2-5 Insertion loss of three inch length of RWG waveguide

Short-Circuited Transmission Line

It is demonstrated in the textbook that the input impedance of a lossless short-circuited transmission line is a pure imaginary function; therefore, the input reactance is given by the following equation.

$$X_{in} = Z_O \tan\theta$$

Where $\theta = \beta d$ is the electrical length of the transmission line in degrees

Therefore, we can see that this reactance can change from inductive to capacitive depending on the length of the transmission line.

Example 2.11-1: In ADS plot the reactance of a loss less short-circuited transmission line as a function of the electrical length of the line.

Solution: To plot the reactance of the short-circuited transmission line in ADS, create a schematic with a grounded transmission line. Make the length of the transmission line a variable with any starting value in degrees. Setup a new S parameter simulation with a single frequency at 1500 MHz.

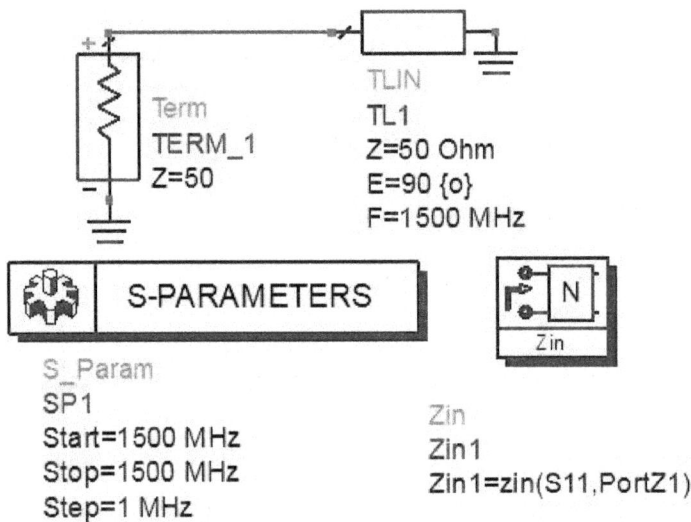

Figure 2-6 Short-circuited line reactance versus electrical length

Then use a Parameter sweep to vary the electrical length of the transmission line from 0 to 360 degrees. Setup a graph to plot the reactance of the shorted transmission line vs. the electrical length from the Parameter Sweep data set as shown in Fig. 2-7.

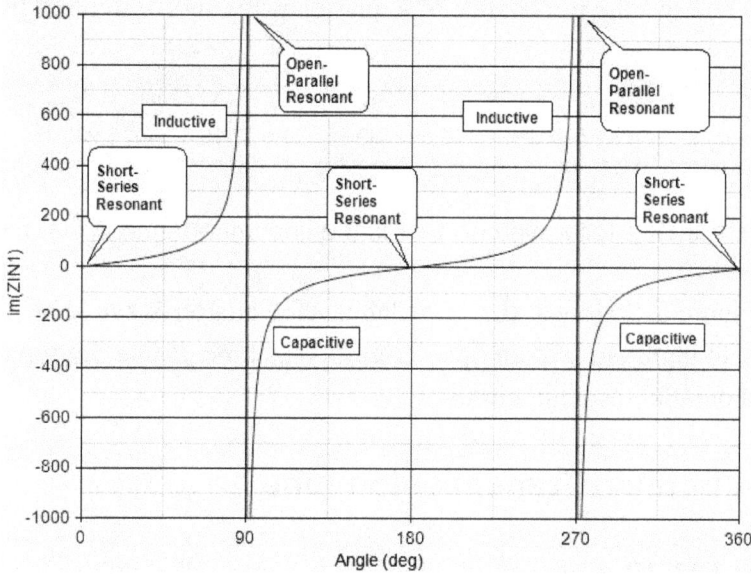

Figure 2-7 Short-circuited line reactance versus electrical length

Example 2.11-2A: Calculate the input impedance of a short-circuited microstrip transmission line for a given electrical length of the line.

Solution: In ADS simply place an ideal transmission line element on a schematic and enter the desired impedance, electrical length, and frequency. Figure 2-8 shows the correct method of modeling a microstrip short-circuited transmission line with a VIA hole.

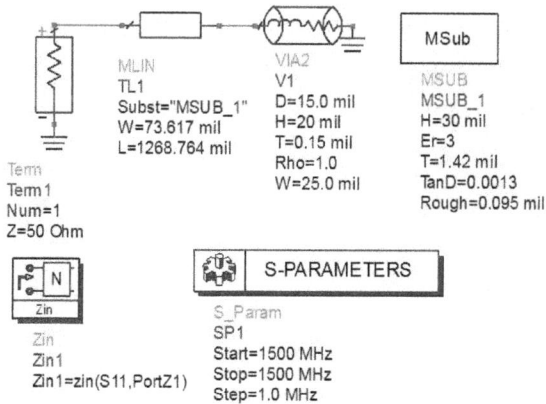

Figure 2-8 Quarter wave short circuited line schematic

Figure 2-9 shows the impedance of a quarter-wave short-circuited line

freq	Zin1	Zin1
1.500 GHz	768.602 - j3.067E3	3.161E3 / -75.930

Figure 2-9 Quarter wave short-circuited line impedance in two formats

As the Figure 2-9 shows, the impedance of a quarter-wave short-circuited line is quite high, close to an open circuit. This type of line section could be used as a parallel resonant circuit.

Open-Circuited Transmission Line

Example 2.11-2B: Calculate the input impedance of a quarter wave open-circuited microstrip transmission line for a given length of the line.

Solution: The reactance of a lossless open circuited transmission line is given by the following equation.

$$X_{in} = Z_O \cot\theta$$

Where: θ is the electrical length of the transmission line in degrees.

Use a parameter sweep in ADS to observe the behavior of this reactance as the length of the open circuit transmission line is varied from 0 to 360 degrees. Note that the transmission line is terminated with a 10^6 Ω load to emulate an open circuit termination on the transmission line.

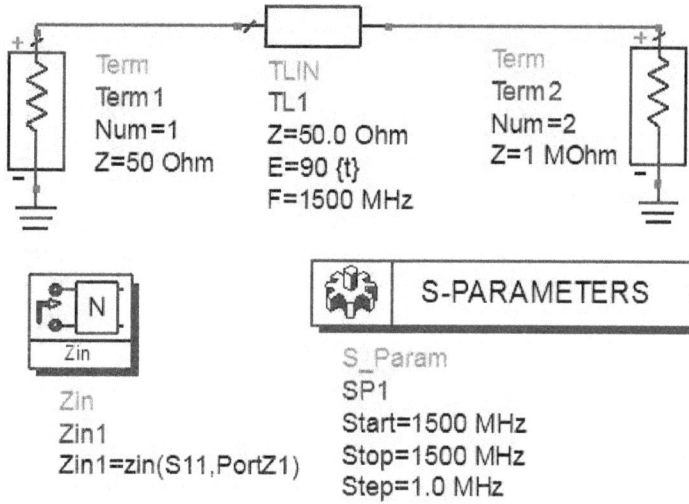

Figure 2-10 Quarter wave Open circuited transmission line schematic

Comparing the open-circuited line reactance to the short-circuited line reactance we can see that a 90°, $\lambda_g/4$, offset is present.

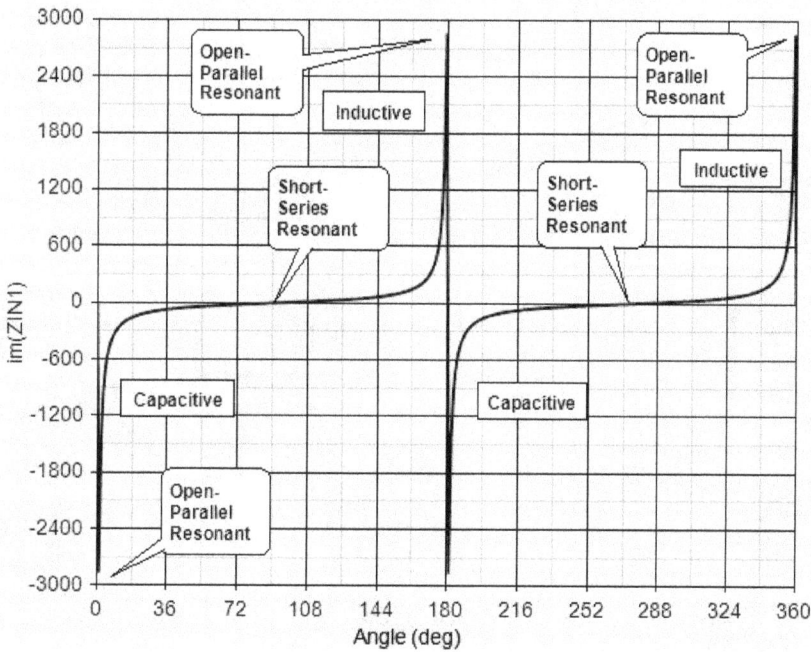

Figure 2-11 Reactance of open circuited line versus electrical length

Modeling Open-Circuited Microstrip Lines

Care must be used when modeling the open circuit microstrip line due to the radiation effects from the end of the transmission line. The E fields that exist in the air space of the microstrip line add capacitance to the microstrip transmission line. On an open circuit microstrip line this fringing capacitance is referred to as an end effect. The end effect makes the line electrically longer than the physical length. This requires that the physical line length be shortened to achieve the desired reactance.

Example 2.11-2C: Calculate the input impedance of a quarter wave open-circuited microstrip transmission line using termination with end effects.

Solution: Figure 2-12 shows the correct method of modeling a microstrip open circuit transmission line using an end effect. As Figure 2-26 shows, the impedance of a quarter-wave section of open circuit line is quite low, close to a short circuit. This type of line section could be used as a series resonant circuit.

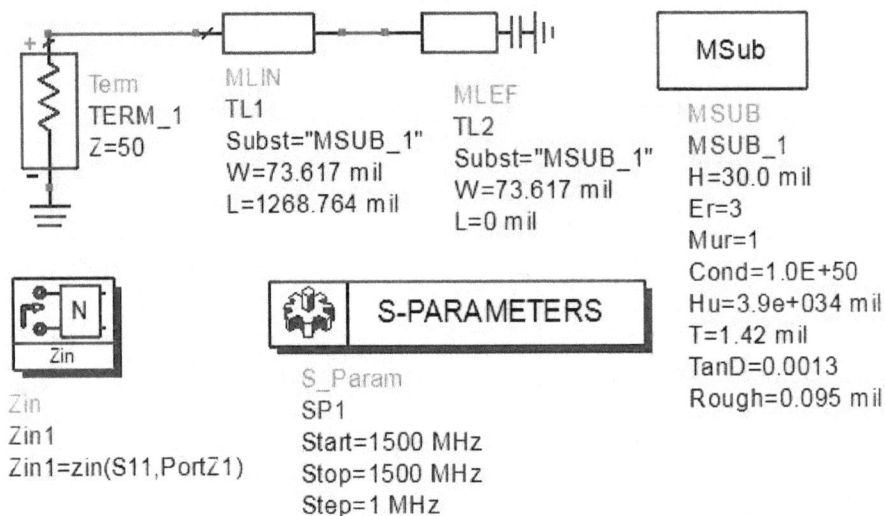

Figure 2-12 Quarter wave open circuited line schematic

Figure 2-13 shows the impedance of a quarter-wave open-circuited line

freq	Zin1	Zin1
1.500 GHz	0.045 + j0.835	0.837 / 86.935

Figure 2-13 Quarter wave open circuited line impedance

Distributed Inductance and Capacitance

For short lengths of high impedance transmission line use the following equations to calculate the length of microstrip line.

$$Inductive\ Line\ \ Length = \frac{f\,\lambda_g\,L}{Z_L}$$

$$Capacitive\ Line\ \ Length = f\,\lambda_g\,Z_C\,C$$

Where:

f = frequency and which inductance is calculated
L= nominal inductance value
C=nominal capacitance value
Z_L= impedance of inductive transmission line
λ_g =wavelength using the effective dielectric constant

Example2.11-2D: Convert the lumped element capacitors and inductors to distributed elements.

Solution: Figure 2-15 shows the low and high impedance microstrip equivalent circuit to the lumped element circuit in Figure 2-14. The PCB layout shows the line width relationship among the microstrip lines.

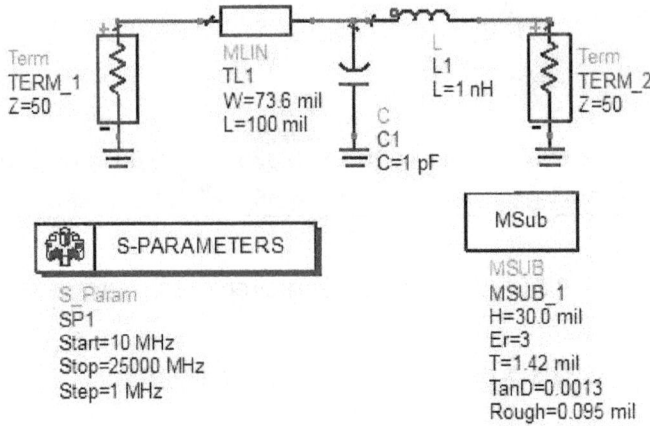

Figure 2-14 Lumped capacitive and inductive lines with PCB layout

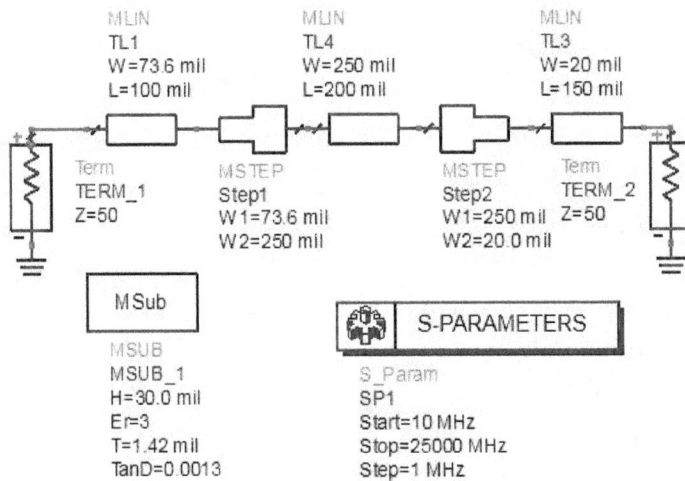

Figure 2-15 Distributed capacitive and inductive lines

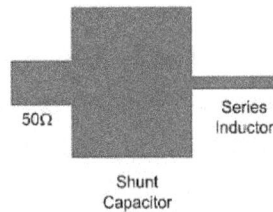

Figure 2-16 Capacitive and inductive lines in PCB layout

2.11.8 Microstrip Bias Feed Networks

Another useful purpose for high impedance and low impedance microstrip transmission lines is the design of bias feed networks. Often it is necessary to insert voltage and current to a device that is attached to a microstrip line. Such a device could be a transistor, MMIC amplifier, or diode. The basic bias feed or "bias decoupling network" consists of an inductor (used as an "RF Choke") and shunt capacitor (bypass capacitor). At lower RF frequencies (< 200 MHz) these networks are almost entirely realized with lumped element components. Even at these low frequencies it is very important to account for the parasitics in the components.

Example 2.11-2E: Design a lumped element biased feed network.

Solution: Fig. 2-17 shows a typical series inductor, shunt capacitor, lumped element bias feed and its effect on a 50 Ω transmission line.

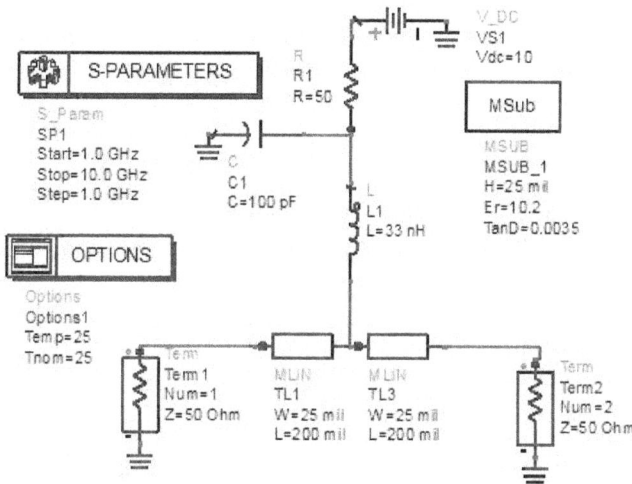

Figure 2-17 Inductor and bypass capacitor bias insertion network

The bias feed response from 0.5 to 5 GHz is shown in Figure 2-18.

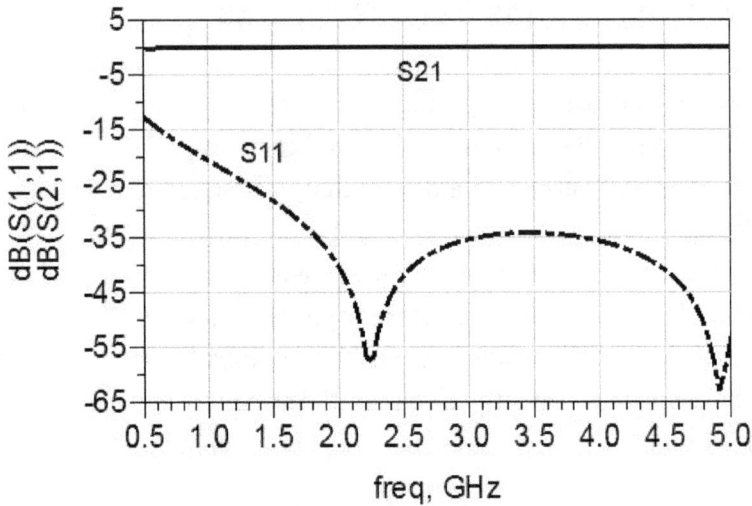

Figure 2-18 Response of the typical bias feed network

Distributed Bias Feed Design

A high impedance microstrip line of $\lambda_g/4$ can be used to replace the lumped element inductor. Similarly a $\lambda_g/4$ of low impedance line can be used to model the shunt capacitor.

Example 2.11-3: Calculate the physical line length of the $\lambda_g/4$ sections of 80 Ω and 20 Ω microstrip lines at a frequency of 2 GHz. Create a schematic of a distributed bias feed network.

Solution: Use the 80 Ω high impedance quarter wave section and a shunt capacitance as shown in Figure 2-19. A microstrip taper, TP1, is used to connect the low impedance line to the high impedance line. Note the use of the microstrip tee junction, TE2. The tee junction accurately models the electrical length of the junction and includes all parasitic effects of the discontinuity. An end-effect element is used on the open circuit line. The response of the bias feed is characterized by the null in the return loss and very low insertion loss near the design frequency of 2 GHz. The return loss null occurs at 1.85 GHz suggesting that the high impedance line length should be decreased to center the design on 2 GHz.

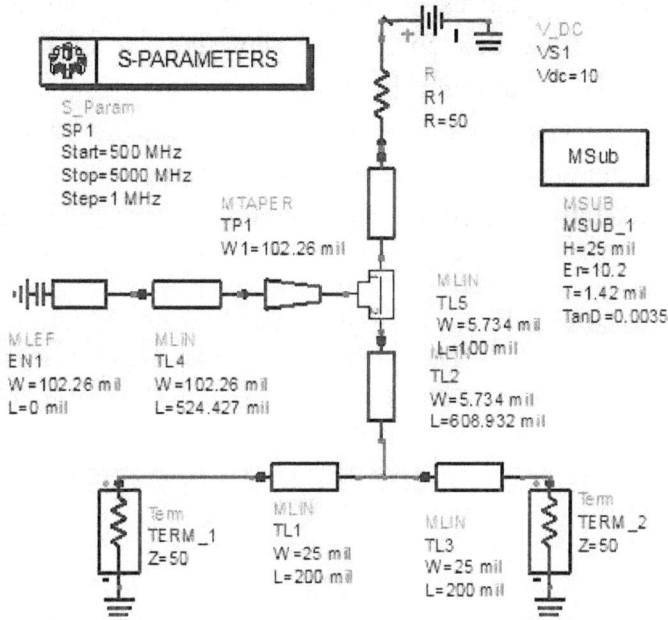

Figure 2-19 Bias feed modeled with distributed transmission line elements

The bias feed response from 0.5 to 5 GHz is shown in Figure 2-20.

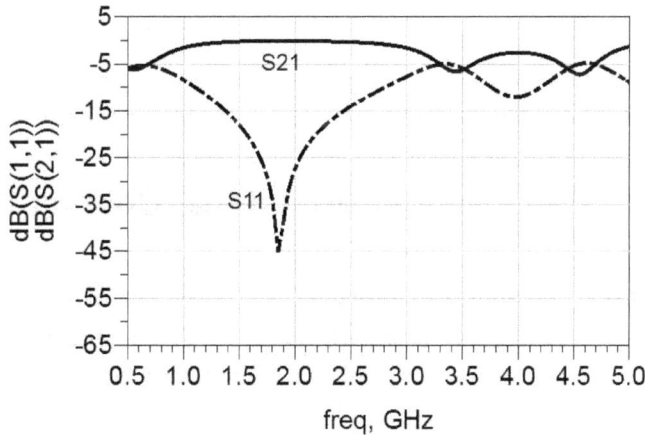

Figure 2-20 Distributed Bias feed response

A modified version of the open circuited transmission line is the radial stub. The radial stub can be used in applications where an open circuit transmission line is needed. Fig. 2-21 shows the use of the radial stub replacing the open circuit transmission line in the bias feed. Comparing the responses we can see that the network using the radial stub achieves a slightly wider bandwidth. This is one of the advantages of using the radial stub. The radial stub may also result in a slightly smaller PCB pattern.

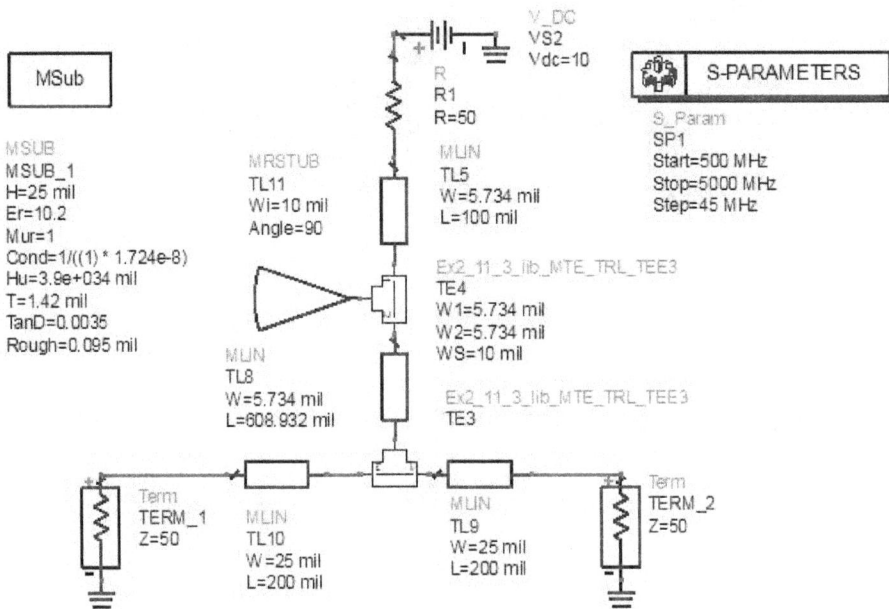

Figure 2-21 Bias feed with open circuited line replaced with the radial stub

Microstrip Edge Coupled Directional Coupler Design

Example 2.12-1: Design a simple edge coupled microstrip directional coupler with a coupling factor of 10 dB at 5 GHz. Use Rogers RO3003 substrate with relative dielectric constant $\varepsilon_r = 3.0$ and 0.020-inch thickness.

Solution: Traditional coupler design required the computation of the even and odd mode impedance based on the characteristic impedance and coupling factor. Many references have tables and families of curves in which the line width and spacing could be obtained. The ADS TLINE utility provides an exact solution for the coupled line parameters. Select the coupled microstrip line calculator from the Rectangular transmission lines in the TLINE utility. Enter the dielectric constant (ε_r), the dielectric thickness (h), and conductor thickness (t) as shown in Figure 2-22. Select the Synthesis mode and choose to synthesize a coupled line pair based on the input characteristic impedance, Z_o, and the coupling in dB. The coupled line width, W, is then calculated be 38.86 mils and the line spacing, s, is 6.16 mils. Add the RO3003, 0.020-inch thick substrate and create the coupler schematic using the Coupled Microstrip Line (Symmetrical) element. On the coupled port side add a Microstrip Bend with Optimal Miter at 90° from the main path. It is important to keep this side orthogonal from the main path so that no further parallel line coupling can occur. The optimal miter element automatically optimizes the miter for the least amount of discontinuity traversing the 90° bend. Lastly add a short section of line to each port to complete the circuit as shown in Figure 2-22. Use a Linear Analysis to sweep the coupler from 4500 MHz to 5500 MHz. Simulate the circuit and display the insertion loss (S21), coupled response (S31), isolation (S41), and the directivity. The directivity must be calculated separately. It is convenient to implement simple mathematical operations directly in the Graph Properties window as opposed to using an Equation block. The directivity is calculated by subtracting the coupled port response from the isolation.

Ex2_12_1_lib_MCP_TRL_COUPLED

Term
TERM_1
Z=50

MLIN
TL1
W=44 mil
L=50 mil

TL2
W=38.86 mil
S=6.16 mil
L=365 mil

MLIN
TL3
W=44 mil
L=50 mil

Term
TERM_2
Z=50

MBEND
Bend1
W=44 mil
Angle=90
M=0.6

MBEND
Bend2
W=44 mil
Angle=90

MLIN
TL4
W=44 mil
L=50 mil

MLIN
TL5
W=44 mil
L=50 mil

M Sub

MSUB
MSUB_1
H=20 mil
Er=3.0
Mur=1
T=1.42 mil
TanD=0.0013
Rough=0.095 mil

Term
TERM_3
Z=50

Term
TERM_4
Z=50

S-PARAMETERS

S_Param
SP1
Start=4500 MHz
Stop=5500 MHz
Step=1 MHz

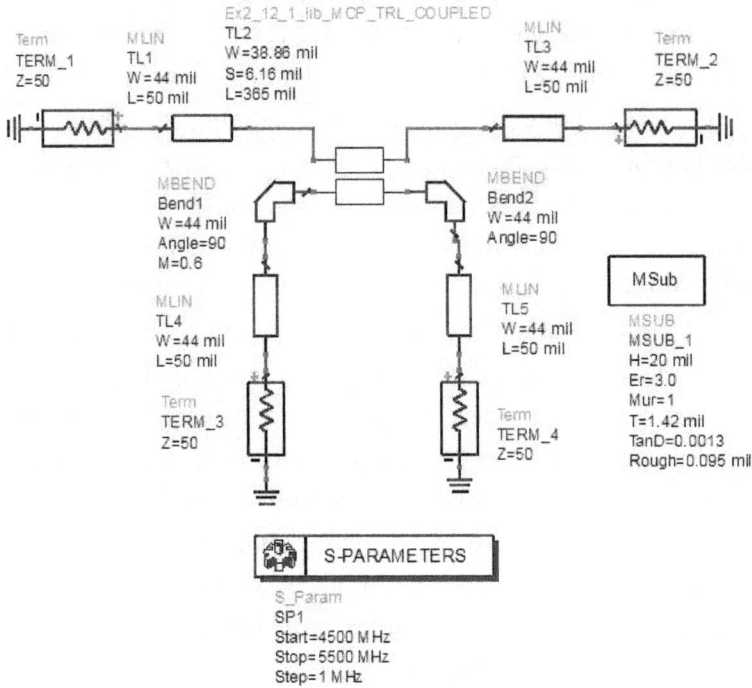

Figure 2-22 Schematic of the microstrip directional coupler

The simulated coupling factor is 10.157 dB which is very close the 10 dB design goal. The isolation is found to be about 25 dB. The directivity of this Directional Coupler is about 15 dB. Clearly this type of directional coupler is not appropriate for VSWR measurement. This type of directional coupler is useful for obtaining a sample of the input signal without disturbing or loading down the input signal. High directivity directional couplers are typically realized with broadside coupled lines in stripline media or in waveguide.

Eqn Directivity=dB(S(4,1)) - dB(S(3,1))

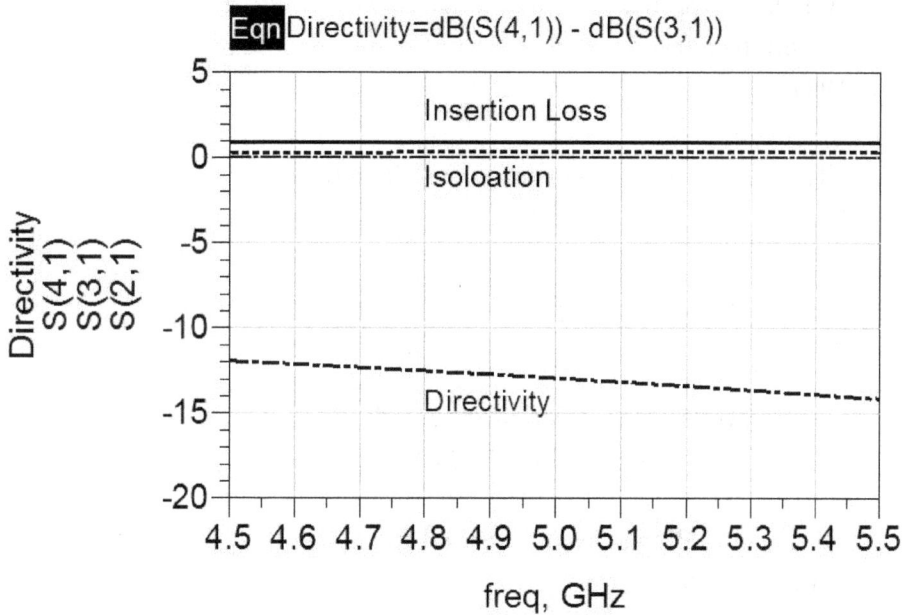

Figure 2-22 Response of the microstrip directional coupler 3

References and Further Reading

[1]   Ali A. Behagi, *RF and Microwave Circuit Design,* A Design Approach Using (**ADS**). Techno Search, Ladera Ranch, CA 92694. August 2015.

[2]   David M. Pozar, *Microwave Engineering*, Third Edition, John Wiley and Sons, Inc. 2005

[3]   Foundations for Microstrip Circuit Design, T.C. Edwards, John Wiley & Sons, New York, 1981

[4]   Keysight Technologies, Manuals for Advanced Design System, *ADS 2016.01 Documentation Set*, Keysight EEsof EDA Division, Santa Rosa, California, www.keysight.com

[5]   William   Sinnema   and   Robert   McPherson,   Electronic
        Communications, Prentice-Hall Canada, Inc., Scarborough, Ontario,
        1991

[6]   UHF/Microwave Experimenters Manual, American Radio Relay
        League, Newington, CT.1990

[7]   Reference: I. J. Bahl and D. K. Trivedi, "A Designer's Guide to
        Microstrip Line", Microwaves, May 1977, pp. 174-182.

[8]   Microwave Handbook Volume 1, Radio Society of Great Britain,
        The Bath Press, Bath, U.K., 1989.

[9]   Tatsuo Itoh, Planar Transmission Line Structures, IEEE Press, New
        York, NY, 1987

Problems

2-1.   Determine the VSWR of a satellite antenna with a return loss of -
        11.4 dB.

2-2.   The input reflection coefficient of a transistor is measured to be 0.22
        at an angle of 32°. Determine the input VSWR of the device.

2-3.   Determine the impedance of a quarter-wave transformer to match a
        25 Ω load to a 50 Ω source.

2-4.   Design the quarter-wave transformer from Problem 3 using a
        microstrip transmission line. The frequency of operation is 2.05
        GHz. The dielectric constant is 3.0 with a thickness of 0.030 in.
        Determine the length and width of the microstrip line.

2-5.   A radio transmitter is operating into a transmission line that
        measures a 3:1 VSWR. Determine the percentage of power that
        would be expected to reflect back into the transmitter.

2-6.   A series RLC load, R = 75 Ω, L = 10 nH, C = 25 pF is connected to a 50 Ω transmission line. Setup a linear analysis in ADS to sweep the frequency from 200 MHz to 2000 MHz in 200 MHz steps. Display the input reflection Coefficient, S11, and VSWR in a Table.

2-7.   Create a simple schematic using the RG8 coaxial cable. Set the length to 50 ft. Calculate the insertion loss in a Table. Terminate the coaxial line with a 100 Ω resistor and display the input return loss and reflection coefficient in the same Table.

2-8.   Calculate the cutoff frequency of the TE1,0 mode in a rectangular waveguide with a height of 0.200 inches and a width of 0.470 inches. Also calculate the waveguide wavelength, λ_g.

2-9.   Design a distributed bias feed network for a C Band amplifier operating at 6.0 GHz. Use a microstrip substrate with a dielectric constant of 10.2 and a thickness of 0.025 inches. Plot the insertion loss and return loss from 2 GHz to 10 GHz.

2-10.   Determine the physical length of a $\lambda_g/4$ open circuit microstrip transmission line with an impedance of 20 Ω. The frequency of operation is 10 GHz. Use a microstrip dielectric constant of 2.2 and a thickness of 0.010 inches. Determine whether an end-effect model element should be used.

Chapter 3

Network Parameters and the Smith Chart

Introduction

At low frequencies below the VHF range, the terminal voltages V_1 and V_2 and the terminal currents I_1 and I_2 of a two-terminal network, shown in Fig. 3-1, can be related to each other by a different set of matrix parameters. The most common representations are the impedance matrix (Z parameters), the admittance matrix (Y parameters), the hybrid matrix (h parameters), and the transmission matrix (ABCD parameters).

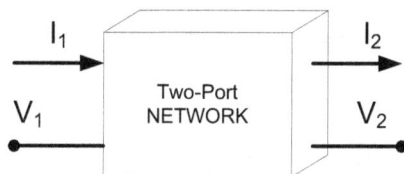

Figure 3-1 Low frequency two-port network

Example 3.4-1: Plot the impedance $Z = 25 + j25\ \Omega$ on the standard Smith Chart.

Solution: To display the impedance on the Smith chart, create a new workspace in ADS and follow the procedure.

1. From the ADS Main window, create a new schematic and Insert S_Prams Template.

2. Connect a wire between the two Terminations and change the load impedance to $25 + j25\ \Omega$.

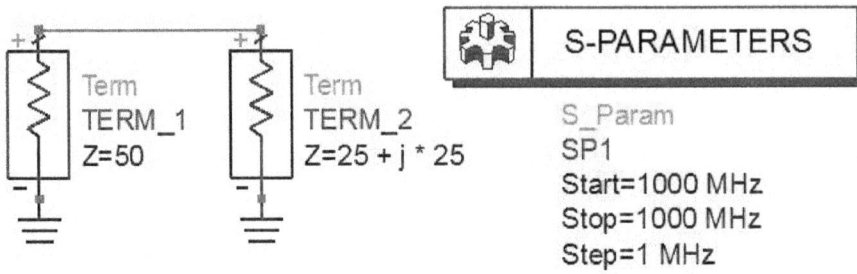

Figure 3-2 Schematic for plotting the impedance on the Smith chart

3. Simulate the schematic and select the Smith chart in data display window.

4. From the list of parameters Add S11 to display the impedance on the Smith chart, as shown in Figure 3-3.

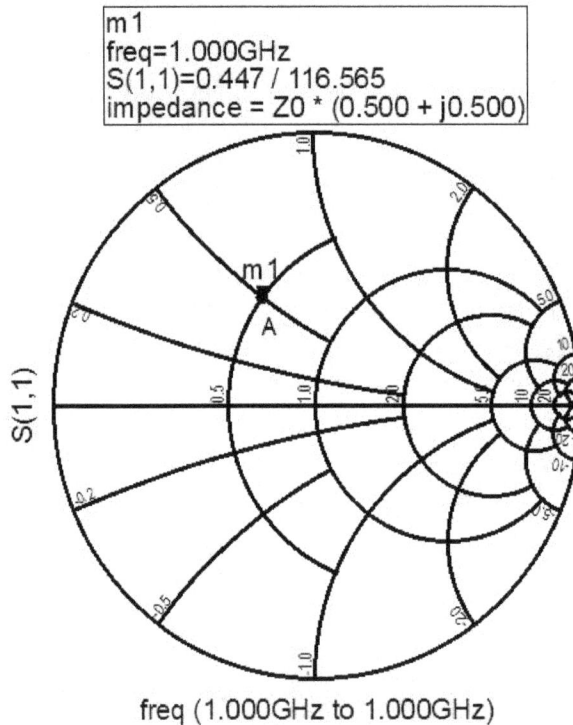

m 1
freq=1.000GHz
S(1,1)=0.447 / 116.565
impedance = Z0 * (0.500 + j0.500)

freq (1.000GHz to 1.000GHz)

Figure 3-3 Plotting impedance on the Smith Chart

Lumped Element Movements on the Smith Chart

Lumped element movements on the Smith Chart form the basis for impedance matching without involving circuit synthesis mathematics.

Example 3.5-1: Measure the amount of reactance required to move the impedance $Z = 25 + j25$ Ω from point A to point B on the Smith Chart, as shown in Figure 3-5.

Solution: The amount of reactance required in the inductor can be measured from the reactance lines that intersect the start point (A) and end point (B). As Figure 3-5 shows the reactance is approximately 0.359 Ω. Using a design frequency of 1000 MHz the inductance is calculated to as 2.86 nH.

$$L\ series = X_n/2PiF = \frac{0.359 \cdot (50)}{2 \cdot \pi \cdot 1000 \cdot 10^6} = 2.86\ nH$$

Add the inductor to the impedance element in ADS to verify that a 2.86 nH inductor moves the impedance to point B on the Smith Chart. Make the inductor tunable and set the initial value to 0.1 nH and tune the value of inductance to move the impedance from point A to point B.

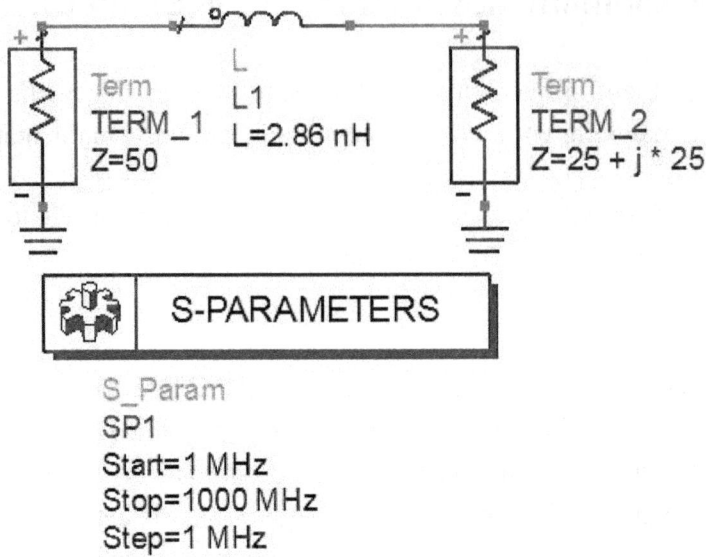

Figure 3-4 Series inductance added to 25 + j25 Ω impedance

Simulate the schematic and display S11 on the Smith chart, as shown in Figure 3-5.

m1
freq=1.000GHz
Design1_TestBench..S(1,1)=0.447 / 116.565
impedance = Z0 * (0.500 + j0.500)

m2
freq=1.000GHz
S(1,1)=0.575 / 90.381
impedance = Z0 * (0.500 + j0.859)

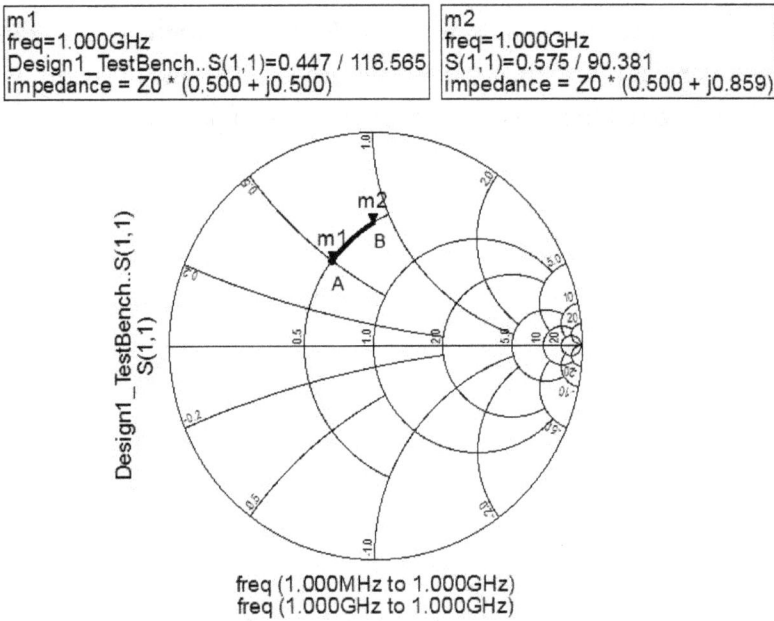

freq (1.000MHz to 1.000GHz)
freq (1.000GHz to 1.000GHz)

Figure 3-5 Moving point A to point B

Adding a Shunt Reactance to an Impedance

Adding a shunt element to an impedance point on the Smith Chart causes the resulting impedance to move along the constant conductance circle in which the impedance intersects. A shunt inductance will move the impedance in a counter-clockwise direction while a shunt capacitance will move the resulting impedance in a clockwise direction on the constant conductance circle. The susceptance that is added to the impedance by the shunt element can be read from the Smith Chart by finding the difference between the lines of susceptance that intersect the start point and end point on the constant conductance circle.

Example 3.5-2A: Measure the amount of susceptance required to move from point B to point C on the real axis.

 Solution:. Add a shunt capacitance to move the impedance from point B to point C as shown in Figure 3-6. When adding a shunt element switch from the impedance grid to the admittance coordinates. The admittance follows the constant conductance circle in which the point lies by the difference between the intersecting susceptance lines. The susceptance as measured on the perimeter of the chart is 0.865 mhos. Calculate the value of capacitance, where, $n = 50$ Ohm.

$$C \ (shunt) = \frac{B}{\omega n} = \frac{0.865}{2 \cdot \pi \cdot 1000 \cdot 10^6 \cdot (50)} = 2.75 \ pF$$

Alternatively you can add a shunt capacitor to the circuit and make the capacitance value tunable. Start with a very low value of approximately 0.1 pf and increase the value of capacitance until the admittance is moved from point B to point C.

Term   INDQ   Term
TERM_1   L1   TERM_2
Z=50   L=2.86 nH   Z=25 + j * 25

C
C1
C=2.75 pF

S-PARAMETERS

S_Param
SP1
Start=1000 MHz
Stop=1000 MHz
Step=1 MHz

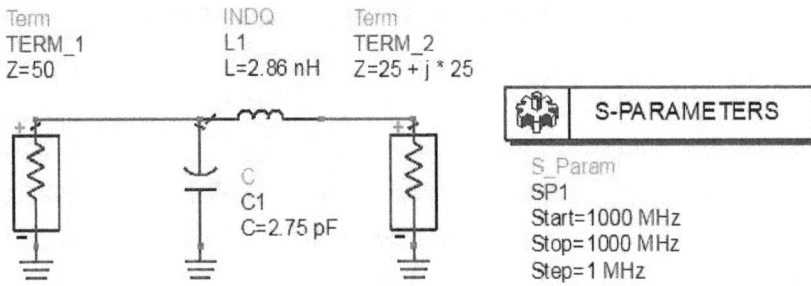

Figure 3-6 Adding the shunt capacitance to the network

Simulate the schematic and display the S11 on the Smith chart, as shown in Figure 3-7.

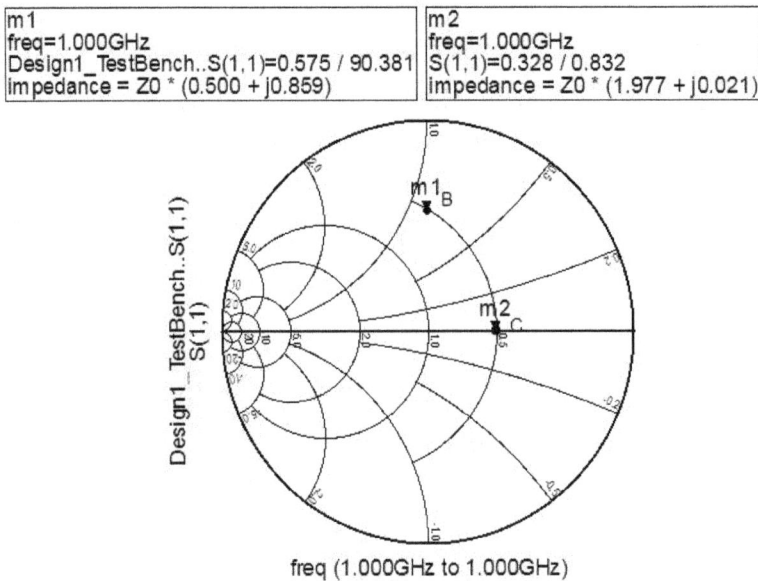

m1
freq=1.000GHz
Design1_TestBench..S(1,1)=0.575 / 90.381
impedance = Z0 * (0.500 + j0.859)

m2
freq=1.000GHz
S(1,1)=0.328 / 0.832
impedance = Z0 * (1.977 + j0.021)

freq (1.000GHz to 1.000GHz)

Figure 3-7 Moving point B to point C

Example 3.5-2B: Show the direction of movement on the Smith chart when adding a series or shunt element to an impedance.

Solution: By iterative addition of series and shunt elements an impedance point can be moved from one location to another on the Smith chart. This process forms the basis of the topic of impedance matching which is fundamental to RF and microwave engineering. ADS workspace Ex3.5-2B demonstrates the step by step procedure as shown in Figure 3-8.

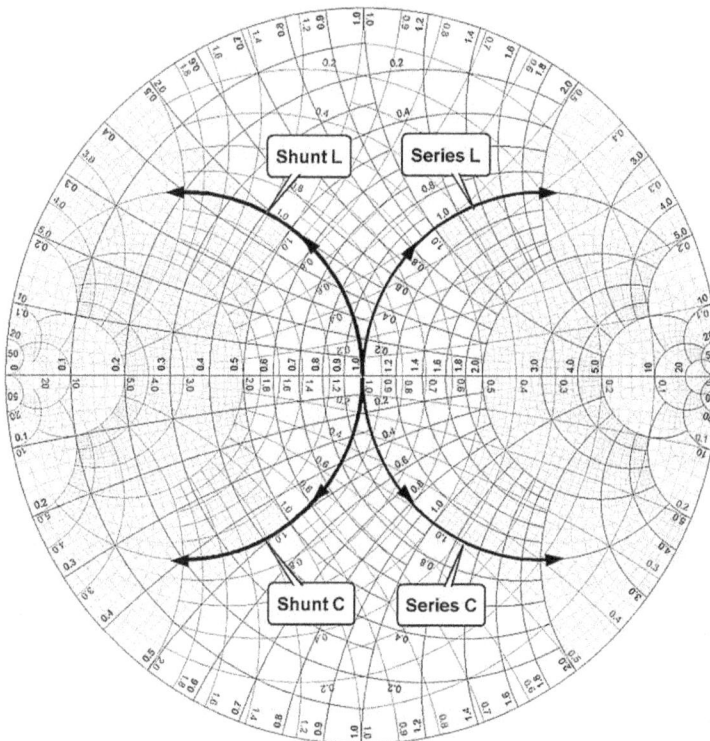

Figure 3-8 Lumped element movements on the Smith chart

VSWR Circles on the Smith Chart

We have shown in the textbook that the VSWR of a network is related to the magnitude of the reflection coefficient, independent of the angle. From plotting the reflection coefficient on the Smith Chart we know that the origin of the impedance vector is located in the center of the chart. This suggests that as the reflection coefficient vector rotates 360 degrees around the chart with a constant magnitude, the VSWR will remain constant. This locus of points around the center of the Smith Chart is known as the constant VSWR circle. In ADS there is no direct way to plot constant VSWR circles, which are frequently used for LNA design. In this example it is shown how to accomplish this task using a few equations.

Figure 3.6-1: Plot the Output VSWR circle on the Smith Chart for VSWR value of 2.25.

Solution: Create a new workspace in ADS and open a new schematic to analyze the network at a fixed frequency as shown in Figure 3-9. Any frequency can be entered as the model uses a fix frequency.

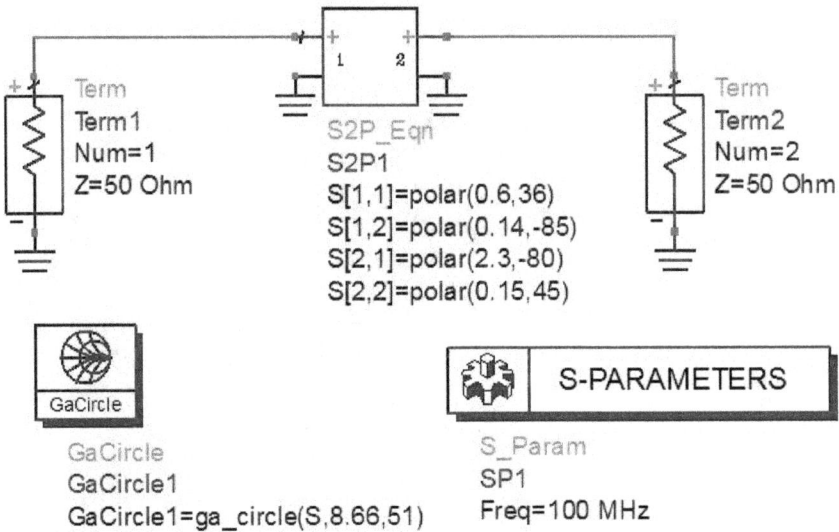

Figure 3-9 S Parameter model in ADS

Enter the desired VSWR value in the Equation and notice the change in the radius of the VSWR circle. In Figure 3-9 a value of 2.25 for VSWR is chosen. Any reflection coefficient intersecting the circle has a VSWR equal to the value of the circle.

If the reflection coefficient is inside the circle, the VSWR is less than the value of the circle. Likewise if a reflection coefficient is outside of the circle, the VSWR is greater than the value of the circle. Figure 3-10 gives intuitive feel for relating reflection coefficients to their VSWR value. From the plot of the VSWR circles in Figure 3-10 we can see that the value of the VSWR is equal to the magnitude of the reflection coefficient as labeled on the horizontal axis to the right of the center of the chart. Conversely the horizontal axis values on the left-hand side of the chart represent $1/\Gamma$.

Eqn gs=polar(0.853,-27.25)   <== Change the value of GammaS here

Eqn VSWRout=2.25   <== Desired VSWR circle here

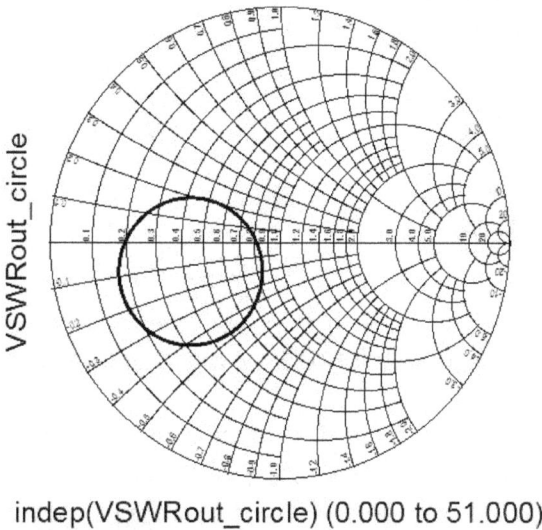

indep(VSWRout_circle) (0.000 to 51.000)

Figure 3-10 Output VSWR circle shown for VSWR value of 2.25

Adding a Transmission Line in Series with an Impedance

We have seen that adding a reactance in series with an impedance point causes the impedance to follow the constant resistance circles. Adding a transmission line of the same impedance as the Smith Chart's normalized impedance, in series with an impedance point causes the resulting impedance to follow the constant VSWR circle in which the impedance lies. The impedance moves in a clockwise direction on the constant VSWR circle.

Example 3.7-1: Revisiting the Z = 25+j25 Ω impedance in ADS, calculate the electrical length of a series transmission line moving the impedance at point A to point B on the Smith Chart, as shown in Figure 3-13.

Solution: Add a series transmission line to the impedance and make the electrical length tunable. Increase the line length to move the impedance to point B. The electrical line length is 58.45 degrees. The impedance on the real axis, on the right hand side of the Smith Chart, would represent a point of maximum voltage-minimum current along the transmission line.

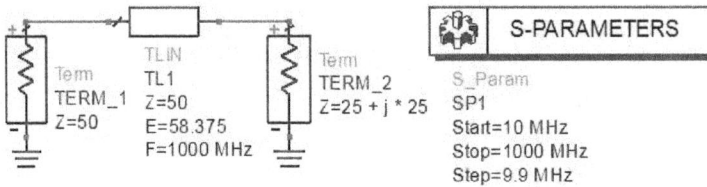

Figure 3-11 Series transmission line added to impedance

Transmission line lengths are sometimes referred to in terms of fractional wavelengths. Because one wavelength is equal to 360°, a 58.45° electrical length represents (58.45/360) 0.162λ. Continue to add electrical length to the transmission line to reach point C on the Smith Chart. As Figure 3-13 shows, the real impedance (zero reactance) on the left side of the horizontal axis on the Smith Chart represents a minimum voltage-maximum current point along the transmission line.

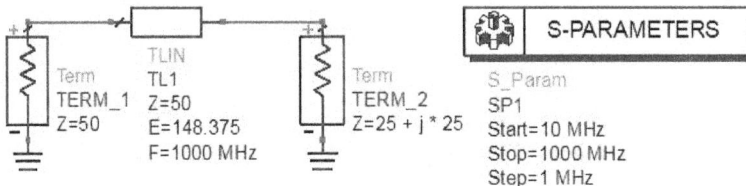

Figure 3-12 Series transmission line added to impedance

m1
freq=10.00MHz
Design1_TestBench..S(1,1)=0.447 / 115.398
impedance = Z0 * (0.505 + j0.510)

m2
freq=1.000GHz
Design1_TestBench..S(1,1)=0.447 / -0.185
impedance = Z0 * (2.618 - j0.009)

m3
freq=1.000GHz
S(1,1)=0.447 / 179.815
impedance = Z0 * (0.382 + j0.001)

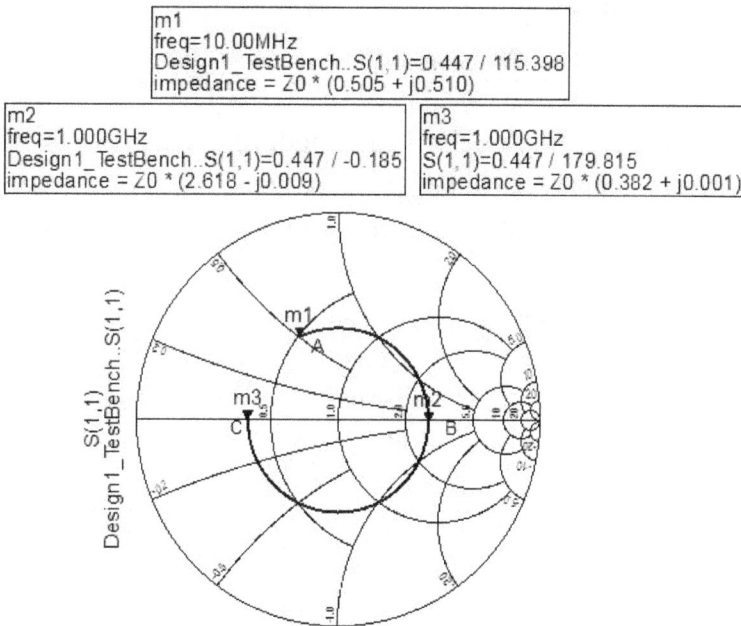

Figure 3-13 Series transmission line moves impedance to a minimum voltage point

Further increasing the length of the transmission line we find that we arrive back at point A at 180 degrees of electrical length. Therefore the electrical distance around the Smith Chart is 180° or λ/2 wavelength. The points of maximum voltage and minimum voltage will repeat every λ/2 wavelength.

Adding a Transmission Line in Parallel with an Impedance

In Chapter 2 we have seen that the open and short-circuited transmission lines could take on the equivalence of an inductor, capacitor, or series and parallel resonant circuits depending on the electrical length of the line. Therefore, the shunt transmission line will behave more like the lumped element movements on the Smith Chart.

Open and Short Circuit Shunt Transmission Lines

For small fractional wavelength transmission lines the open circuit shunt transmission line acts as a shunt capacitor.

Example 3.9-1: Measure the electrical length of a shunt transmission line or the amount of shunt capacitance to move the impedance Z = 25 + j25 Ω from point A to the center of Smith Chart, as shown in Figure 3-14.

Solution: Plotting the impedance on the Smith Chart with the admittance circles shows that the impedance lies directly on the unit conductance circle. Therefore, an open circuit shunt transmission line can move this impedance directly to 50 Ω. As the schematic of Figure 3-14 shows, a 45.38° electrical length of a 50 Ω shunt transmission line moves the impedance to the center of the Smith Chart.

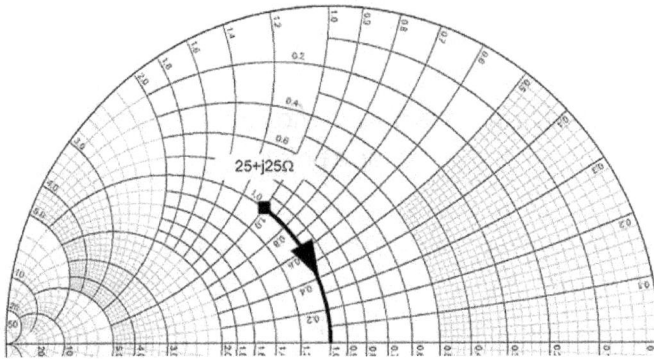

Figure 3-14 Open circuit shunt transmission line added to 25+j25 Ω impedance

For small fractional wavelength transmission lines the short circuit shunt transmission line acts as a shunt inductor. Consider the 4.3 − j14 Ω impedance as shown in Figure 3-14. This impedance lies on the unit conductance circle on the bottom half of the Smith Chart. A 50 Ω shunt transmission line added to the impedance moves along the constant conductance circle to the center of the Smith Chart. Similarly a 2.45 nH shunt inductor has the same effect at a frequency of 1000 MHz. These movements form the basis for distributed network impedance matching which is covered in detail in Chapter 6.

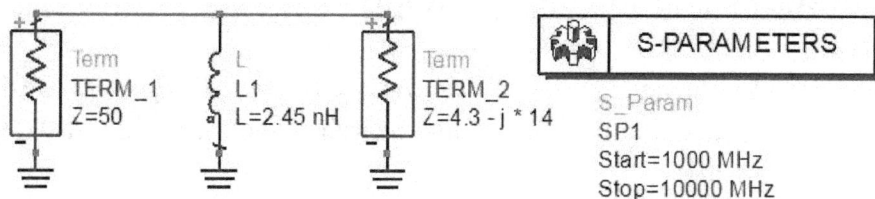

Figure 3-15 Short circuit shunt transmission line added to an impedance

Figure 3-16 shows the movement to the center of Smith chart.

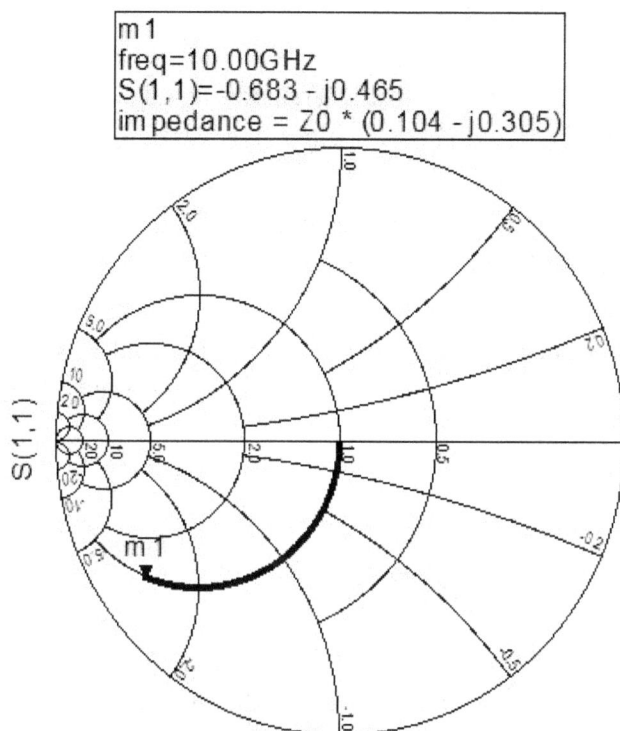

```
m1
freq=10.00GHz
S(1,1)=-0.683 - j0.465
impedance = Z0 * (0.104 - j0.305)
```

Figure 3-16 Short circuit shunt transmission line added to 4.3-j14 Ω
impedance

References and Further Reading

[1]   Ali A. Behagi, *RF and Microwave Circuit Design,* A Design Approach Using (**ADS**). Techno Search, Ladera Ranch, CA 92694. August 2015.

[2]   David M. Pozar, *Microwave Engineering*, Third Edition, John Wiley and Sons, Inc., 2005.

[3]   Franklin F. Kuo, Network Analysis and Synthesis, John Wiley and Sons Inc., 1966

[4]   William Sinnema, Electronic Transmission Technology, Prentice-Hall, Inc, Englewood Cliffs, New Jersey 07632, 1979

[5]   Chris Bowick, RF Circuit Design, Second Edition, Newnes, Elsevier, 2008

[6]   Keysight Technologies, Manuals for Advanced Design System, *ADS 2016.01 Documentation Set*, Keysight EEsof EDA Division, Santa Rosa, California, www.keysight.com

Problems

3-1.   Place a 20 + j30 Ω impedance at point A on the Smith Chart. Add a series inductance to move the impedance along the constant resistance circle to point B having the impedance 20 + j50 Ω. Using a design frequency of 1000MHz, calculate the inductance that this reactance represents.

3-2.   Continuing with Problem 3-1, enable the admittance coordinates on the Smith Chart to add a shunt element. Add a shunt capacitance to move the impedance to the real axis. Measure the susceptance required to move to the real impedance axis by the difference between the intersecting susceptance lines.

3-3.   Calculate the magnitude of reflection coefficient for a desired VSWR = 2.

3-4.   Using ADS, create a constant VSWR circle for a VSWR=20. Comment on the VSWR value required to place the VSWR circle on the circumference of the Smith Chart.

3-5.   A load of 75 + j20 Ω is connected to a 50 Ω transmission line. Calculate the load admittance and the input impedance if the line is 0.2 wavelengths long.

3-6.   For the load impedance in Problem 3-4, determine the reflection coefficient and the transmission coefficient.

3-7.   For the load impedance in Problem 3-4, determine the normalized value of the load impedance if the impedance is normalized to 75 Ω.

3-8.   A series RLC load, R = 100 Ω, L = 20 nH, C = 25 pF is connected to a 50 Ω transmission line. Calculate the VSWR and reflection coefficient at the load at 100 MHz.

3-9.   Using the RLC load impedance of problem 3-8 determine the impedance with a series transmission line of characteristic impedance of 50 Ω and electrical length of 180 degrees.

3-10. Determine the input impedance of a network that has a reflection coefficient of 0.5 at an angle of 112°.

3-11. Create a one-port S parameter text file with the impedance of Problem 3-10 at a frequency of 1 GHz. Plot the S parameter, S11, on the Smith Chart.

3-12. Using the S parameter file of the C Band amplifier, plot the input return loss, S11, and output return loss, S22, on the Smith Chart. Determine the worst case input and output VSWR for this amplifier.

Chapter 4

Resonant Circuits and Filters

Series Resonant Circuits

Example 4.2-1: Consider the one port series resonator that is represented as a series RLC circuit of Figure 4-1. Analyze the circuit, with R = 10 Ω, L = 10 nH, and C = 10 pF.

Solution: The schematic for the analysis in ADS is as follows.

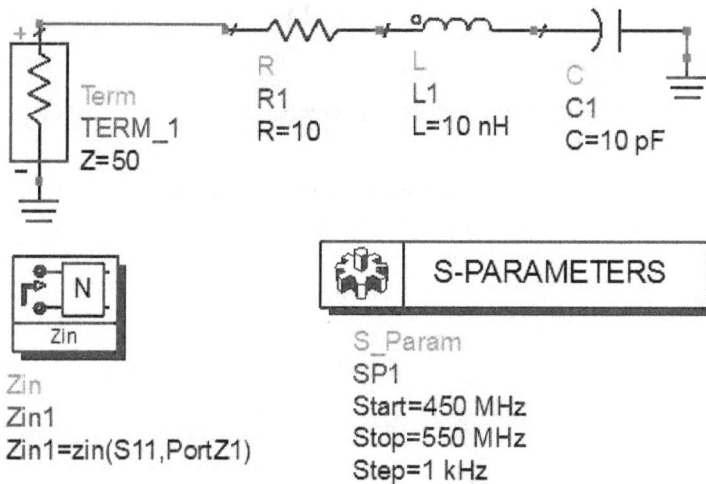

Figure 4-1 One-port series RLC resonator circuit

The plot of the resonator's input impedance in Figure 4-2 shows that the resonance frequency is about 503.3 MHz and the input impedance at resonance is 10 Ω, the value of the resistor in the network.

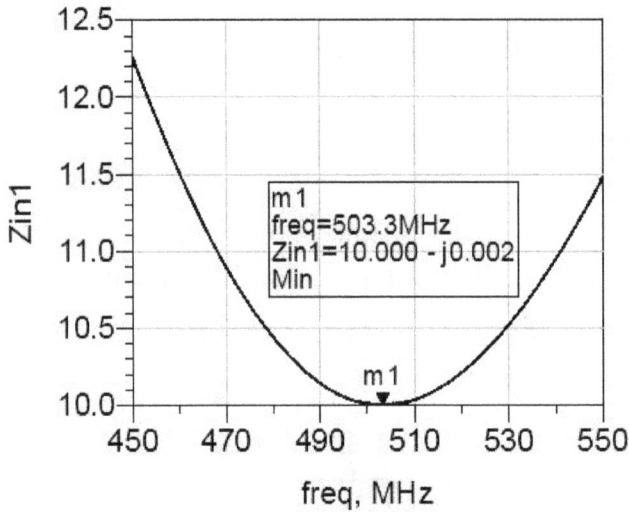

Figure 4-2 Input impedance plot showing the resonance frequency

The input impedance of the series RLC resonant circuit is given by,

$$Z_{in} = R + j\omega L - j\frac{1}{\omega C}$$

Where, $\omega = 2\pi f$ is the angular frequency in radian per second.

If the AC current flowing in the series resonant circuit is I, then the complex power delivered to the resonator is

$$P_{in} = \frac{|I|^2}{2} Z_{in} \quad \frac{|I|^2}{2}\left(R + j\omega L - j\frac{1}{\omega C}\right)$$

At resonance the reactive power of the inductor is equal to the reactive power of the capacitor. Therefore, the power delivered to the resonator is equal to the power dissipated in the resistor

$$P_{in} = \frac{|I|^2 R}{2}$$

Parallel Resonant Circuits

Example 4.2-2: Analyze a rearrangement of the RLC components of Figure 4-1 into the parallel configuration of Figure 4-3. The schematic of Figure 4-3 represents the lumped element representation of the parallel resonant circuit.

Solution: The one port parallel resonant circuit is shown in Figure 4-3.

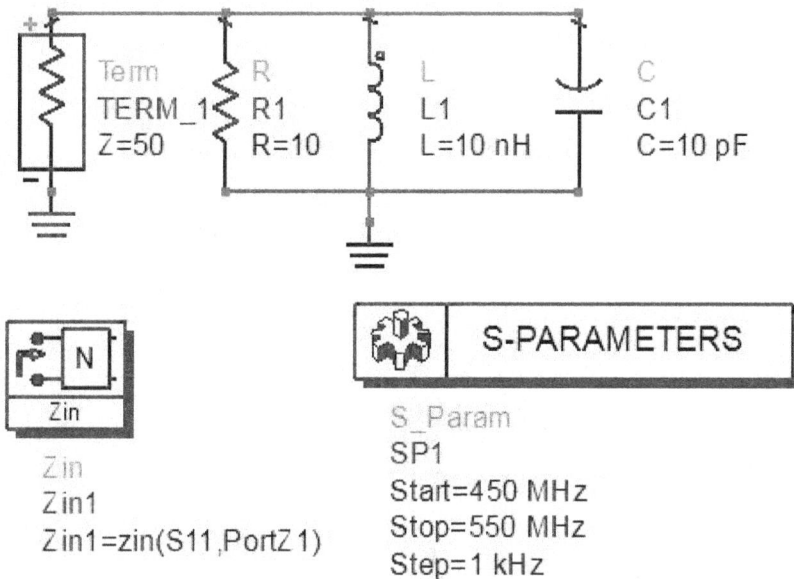

Figure 4-3 One-port parallel RLC resonant circuit

Simulate the schematic and display the input impedance in a rectangular plot. The plot of the magnitude of the input impedance shows that the resonance frequency is still 503.3 MHz where the input impedance is R = 10 Ω. Again this shows that the impedance of the inductor cancels the impedance of the capacitor at resonance. In other words, the reactance, X_L, is equal to the reactance, X_C, at the resonance frequency.

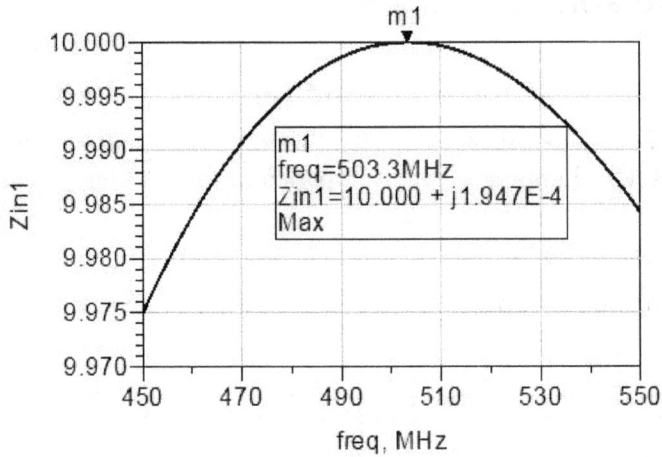

Figure 4-4 Input impedance of parallel RLC resonant circuit

The input admittance of the parallel resonant circuit is given by:

$$Y_{IN} = \frac{1}{R} + j\omega C - j\frac{1}{\omega L}$$

If the AC voltage across the parallel resonant circuit is V, then the complex power delivered to the resonator is

$$P_{in} = \frac{|V|^2}{2}Y_{in} \quad \frac{|V|^2}{2}\left(\frac{1}{R} + j\omega C - j\frac{1}{\omega L}\right)$$

At resonance the reactive power of the inductor is equal to the reactive power of the capacitor. Therefore, the power delivered to the resonator is equal to the power dissipated in the resistor

$$P_{in} = \frac{|V|^2}{2R}$$

The resonance frequency for the parallel resonant circuit as well as the series resonant circuit is obtained by setting $\omega_0 C = \frac{1}{\omega_0 L}$ or:

$$\omega_o = 2\pi f_o = \frac{1}{\sqrt{LC}}$$

Where, ω_0 is the angular frequency in radian per second and f_0 is equal to the frequency in Hertz.

4.2.3 Resonant Circuit Loss

In Figure 4-1 and 4-3 the resistor R1 represents the loss in the resonator. It includes the losses in the capacitor as well as the inductor. The Q factor can be shown to be a ratio of the energy stored in the inductor and capacitor to the power dissipated in the resistor as a function of frequency. For the series resonant circuit of Figure 4-1 the unloaded Q factor is defined by:

$$Q_u = \frac{X}{R} \quad \frac{\omega_o L}{R} = \frac{1}{\omega_o RC}$$

The unloaded Q factor of the parallel resonant circuit in Figure 4-3 is simply the inverse of the series resonant circuit.

$$Q_u = \frac{R}{X} \quad \frac{R}{\omega_o L} = \omega_o RC$$

We can clearly see that as the resistance increases in the series resonant circuit, the Q factor decreases. Conversely as the resistance increases in the parallel resonant circuit, the Q factor increases.

The Q factor is a measure of loss in the resonant circuit. Thus a higher Q corresponds to lower loss and a lower Q corresponds to a higher loss. It is usually desirable to achieve high Q factors in a resonator as it will lead to lower losses in filter applications or lower phase noise in oscillators.

Note that the resonator Q is defined as Q_u, the unloaded Q of the resonator. This means that the resonator is not connected

Equations would then have to be modified to include the source and load resistance. We might also surmise that any reactance associated with the source or load impedance may alter the resonant frequency of the resonator. This leads to two additional definitions of Q factor that the engineer must consider: the loaded Q and external Q.

Loaded Q and External Q

Example 4.2-3: Analyze the parallel resonator that is attached to a 50 Ω source and load as shown in Figure 4-5.

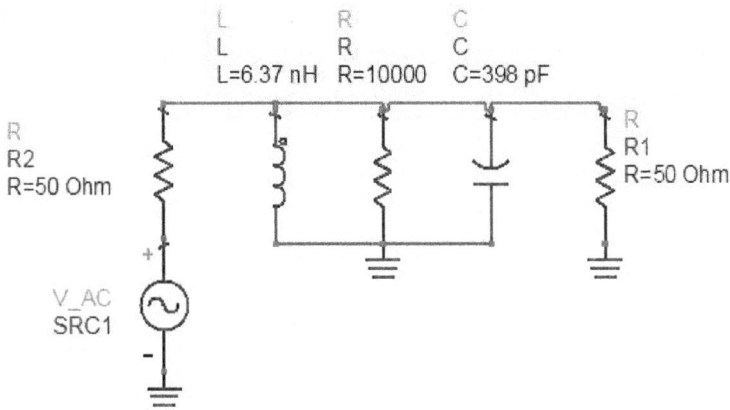

Figure 4-5 Parallel resonator with source and load impedance attached

Solution: Using Equation (4-7) to define the Q factor for the circuit requires that we include the source and load resistance which is 'loading' the resonator. This leads to the definition of the loaded Q, Q$_L$, for the parallel resonator as defined by:

$$Q_L = \frac{R_S + R + R_L}{\omega_o L}$$

Conversely we can define a Q factor in terms of only the external source and load resistance. This leads to the definition of the external Q, Q$_E$.

$$Q_E = \frac{R_S + R_L}{\omega_o L}$$

The Q factors are related by the inverse relationship.

$$\frac{1}{Q_L} = \frac{1}{Q_E} + \frac{1}{Q_U}$$

Lumped Element Parallel Resonator Design

Example 4.3-1: In this example we design a lumped element parallel resonator at a frequency of 100 MHz. The resonator is intended to operate between a source resistance of 100 Ω and a load resistance of 400 Ω.

Solution: Best accuracy would be obtained by using S parameter files or Modelithics models for the inductor and capacitor. However a good first order model can be obtained by using the ADS inductor and capacitor models that include the component Q factor. These models save us the work of calculating the equivalent resistive part of the inductor and capacitor model. The component Q factors are shown in the schematic of Figure 4-6.

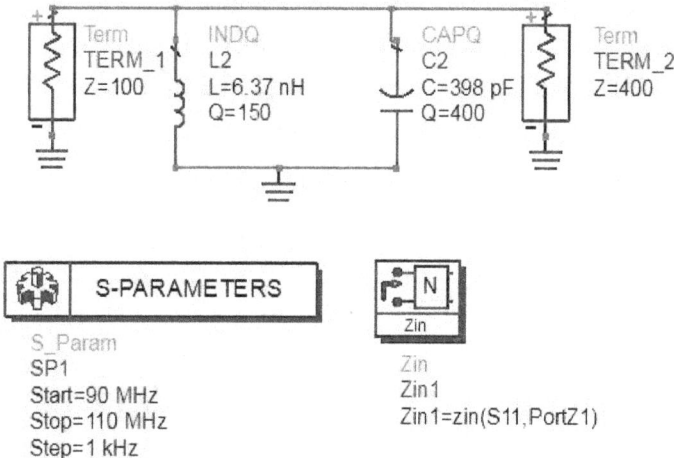

Term
TERM_1
Z=100

INDQ
L2
L=6.37 nH
Q=150

CAPQ
C2
C=398 pF
Q=400

Term
TERM_2
Z=400

S-PARAMETERS

S_Param
SP1
Start=90 MHz
Stop=110 MHz
Step=1 kHz

Zin

Zin
Zin1
Zin1=zin(S11,PortZ1)

Figure 4-6 Parallel resonator using components with assigned Q factors

Simulate the schematic and display the insertion loss in a rectangular plot.

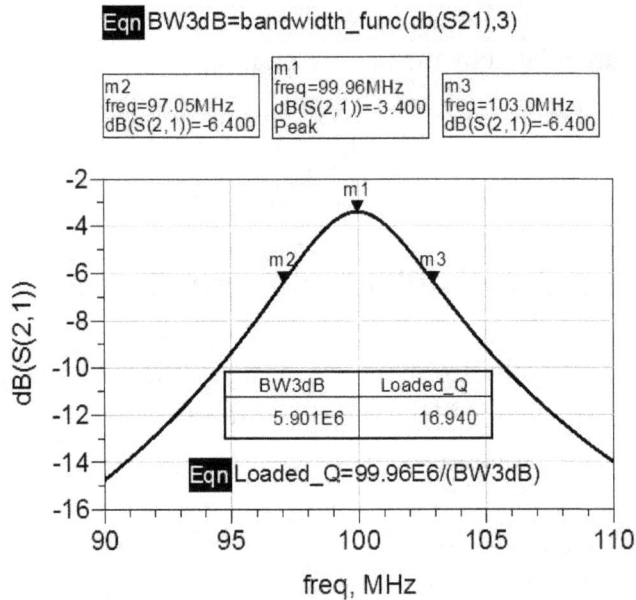

Eqn BW3dB=bandwidth_func(db(S21),3)

m2
freq=97.05MHz
dB(S(2,1))=-6.400

m1
freq=99.96MHz
dB(S(2,1))=-3.400
Peak

m3
freq=103.0MHz
dB(S(2,1))=-6.400

BW3dB	Loaded_Q
5.901E6	16.940

Eqn Loaded_Q=99.96E6/(BW3dB)

Figure 4-7 Insertion loss and calculation of 3 dB bandwidth and Q factor

Note that markers have been placed on the plot of the insertion loss, S21, that gives readout of -3 dB bandwidth. To place markers m2 and m3 manually at -3dB points, first place marker m1 at the peak of S21 trace. Next place marker m2 at a point below marker m1. Select marker m2 and press the up and down arrows to move the marker m2 as close to -3dB point as possible. Repeat the process for marker m3. The loaded Q is calculated by the following equation.

$$Q_L = \frac{\sqrt{f_l f_h}}{BW} \quad \frac{99.98MHz}{5.901MHz} = 16.94$$

Effect of Load Resistance on Bandwidth and Q_L

In RF circuits and systems the impedances encountered are often quite low, ranging from 1 Ω to 50 Ω. It may not be practical to have a source impedance of 100 Ω and a load impedance of 400 Ω.

Example 4.3-2: In the parallel LC Example 4.3-1, change the load from 400 Ω to 50 Ω and re-examine the circuit's 3 dB bandwidth and Q_L.

Solution: Change the load resistance from 400 Ω to 50 Ω as shown in Figure 4-8.

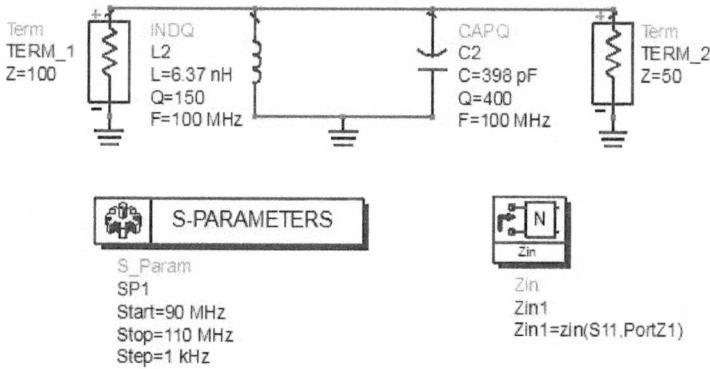

Term
TERM_1
Z=100

INDQ
L2
L=6.37 nH
Q=150
F=100 MHz

CAPQ
C2
C=398 pF
Q=400
F=100 MHz

Term
TERM_2
Z=50

S-PARAMETERS

S_Param
SP1
Start=90 MHz
Stop=110 MHz
Step=1 kHz

N

Zin
Zin1
Zin1=zin(S11.PortZ1)

Figure 4-8 Parallel resonance circuit

Simulate the schematic and display the insertion loss, S21, in a rectangular plot.

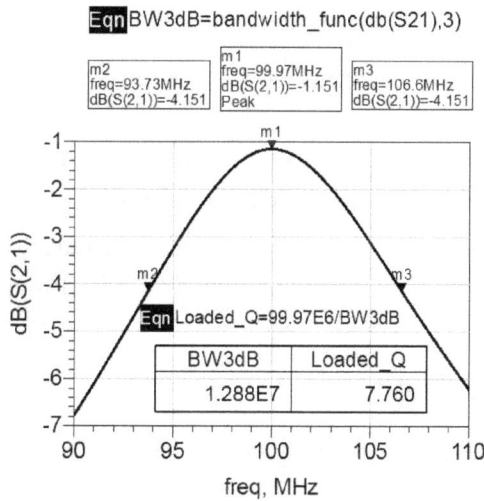

Eqn BW3dB=bandwidth_func(db(S21),3)

m2
freq=93.73MHz
dB(S(2,1))=-4.151

m1
freq=99.97MHz
dB(S(2,1))=-1.151
Peak

m3
freq=106.6MHz
dB(S(2,1))=-4.151

Eqn Loaded_Q=99.97E6/BW3dB

BW3dB	Loaded_Q
1.288E7	7.760

freq, MHz

Figure 4-9 Insertion loss of parallel resonance circuit

Notice the 3 dB bandwidth is now 12.88 MHz resulting in a loaded Q factor of 7.766.

$$Q_L = \frac{\sqrt{f_l f_h}}{BW} \quad \frac{99.958 MHz}{12.88 MHz} = 7.76$$

The loaded Q factor has decreased by nearly half of the original value. We have increased the bandwidth or de-Q'd the resonator. This can also be thought of as tighter coupling of the resonator to the load.

Tapped Capacitor Resonator Design

Example 4.4-1: Consider rearranging the parallel LC network of Figure 4-8 with the tapped capacitor network shown in Figure 4-10. Re-examine the circuit's 3 dB bandwidth and Q_L.

Solution: The new capacitor values for C1 and C2 can be found by the simultaneous solution of the following equations.

$$C_T = \frac{C1 \cdot C2}{C1 + C2}$$

$$R_{L1} = R_L \left(1 + \frac{C1}{C2}\right)^2$$

R_{L1} is the higher, transformed, load resistance. In this example substitute R_{L1} = 400 Ω, the original load resistance value. C_T is simply the original capacitance of 398 pf. The capacitor values are found to be: C1 = 1126.23 pF and C2 = 616.1 pF. The new resonator circuit is shown in Figure 4-10.

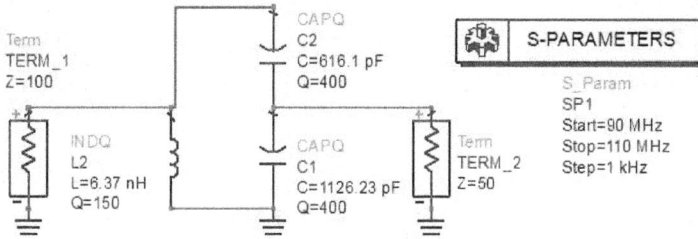

Figure 4-10 Parallel LC resonator using tapped capacitor

Sweeping the circuit we see that the response has returned to the original performance of Figure 4-11.

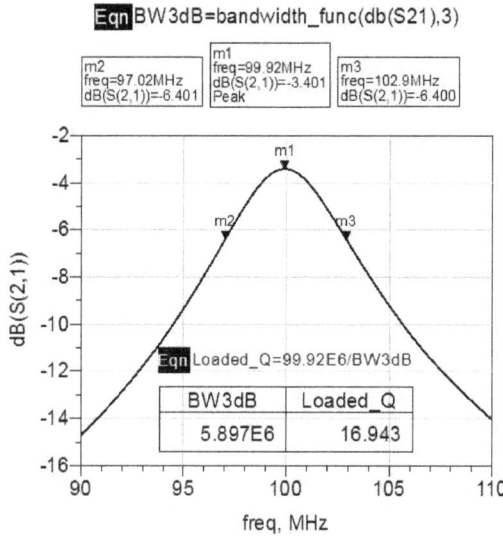

Figure 4-11 Response of parallel LC resonator using a tapped capacitor

The 3 dB bandwidth has returned to 5.89 MHz making the Q_L equal to:

$$Q_L = \frac{\sqrt{f_l f_h}}{BW} = \frac{99.92 MHz}{5.897 MHz} = 16.94$$

The 50 Ω load resistor has been successfully decoupled from the resonator.

The tapped capacitor and inductor resonators are popular methods of decoupling RF and lower microwave frequency resonators. It is frequently seen in RF oscillator topologies such as the Colpitts oscillator in the VHF frequency range.

Tapped Inductor Resonator Design

Example 4.4-2: Similarly design a tapped inductor network to decouple the 50 Ω source impedance from loading the resonator.

Solution: Replace the 100 Ω source impedance with a 50 Ω source and use a tapped inductor network to transform the new 50 Ω source to 100 Ω. Modify the circuit to split the 6.37 nH inductor, L_T, into two series inductors, L1 and L2. The inductor values can then be calculated by solving the following equation set simultaneously. R_{S1} is the higher, transformed, source resistance. In this example substitute $R_{S1} = 100$ Ω,

$$Rs1 = Rs\left(\frac{L_T}{L_1}\right)^2$$

$$L_T = L_1 + L_2$$

Solving the equation set results in values of L1=4.5 nH and L2=1.87 nH. The resulting schematic and response is shown in Figure 4-12.

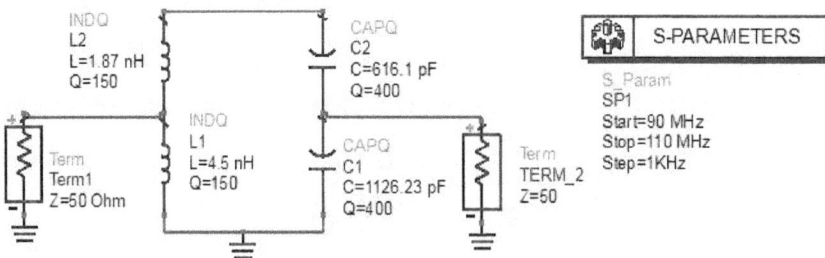

Figure 4-12 Tapped-inductor parallel resonant circuit

The new response is identical to the plot of Figure 4-9. Therefore we now have a source and load resistance of 50 Ω and have not reduced the Q of the resonator from what we had with the original source resistance of 100 Ω and a load resistance of 400 Ω.

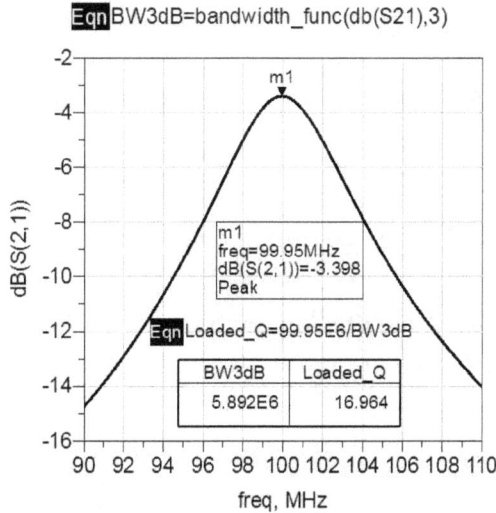

Figure 4-13 Simulated response of the parallel resonant circuit

ADS Model of the Microstrip Resonator

The half wave open-circuited microstrip resonator is modeled in ADS as shown in Fig 4-14. Note that the source and load impedance has been increased to 5000 Ω to avoid loading the impedance of the parallel resonant circuit.

Example 4.5-1: Perform a linear sweep of the resonator using 0.1 MHz step from 4500 MHz to 5400 MHz.

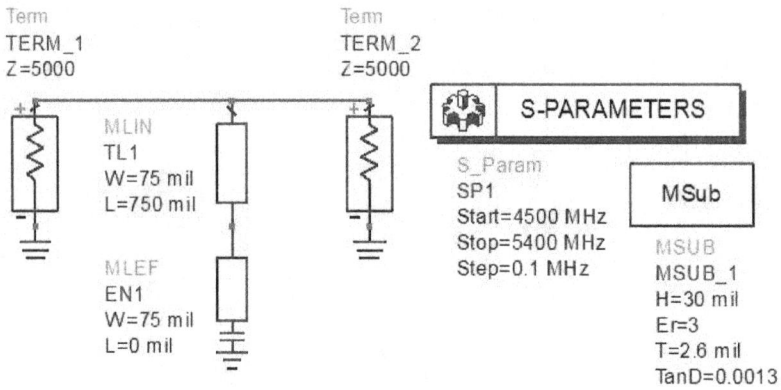

Figure 4-14 Half-wave open ended microstrip resonator

Simulate the schematic and display the S21, the -3 dB bandwidth, and the loaded Q factor, as shown in Figure 4-15.

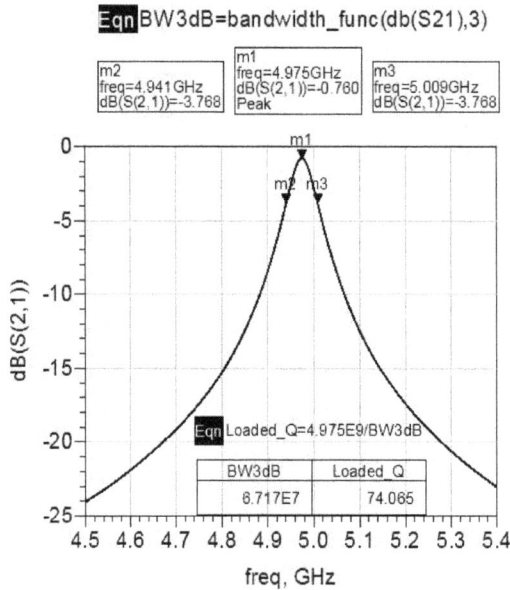

Figure 4-15 Response of the half-wave open circuit microstrip resonator

Using the techniques of section 4.2.3 and Equation (4-10, the 3dB bandwidth is measured to determine the loaded Q of the resonator.

$$Q_L = \frac{4975MHz}{67MHz} = 74.25$$

The insertion loss at the resonant frequency can be used to relate the loaded Q factor to the Q_u.

$$InsertionLoss(dB) = 20\log\frac{Q_u}{Q_u - Q_L}$$

Or:

$$0.76 = 20\log\frac{Q_u}{Q_u - 74.25}$$

$$Q_u = 886$$

Resonator Series Reactance Coupling

Example 4.6-1: As Figure 4-17 shows, the value of the coupling capacitor also has an impact on the size of the Q circle. The diameter of the Q circle is dependent on the coupling of the resonator to the 50 Ω source.

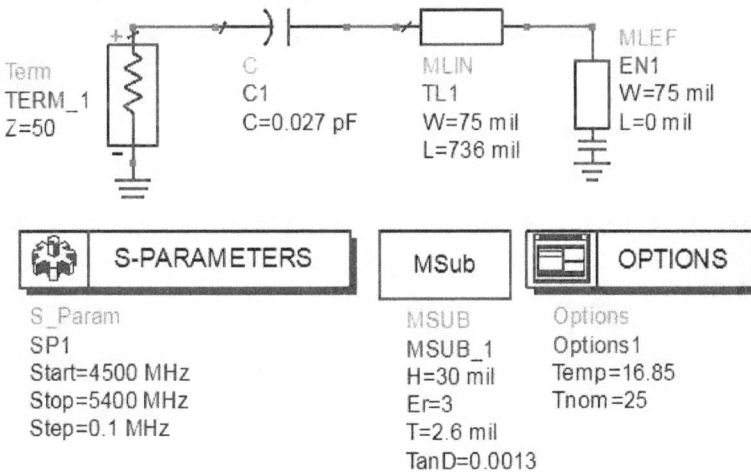

Figure 4-16 Capacitive coupled half-wave microstrip resonator

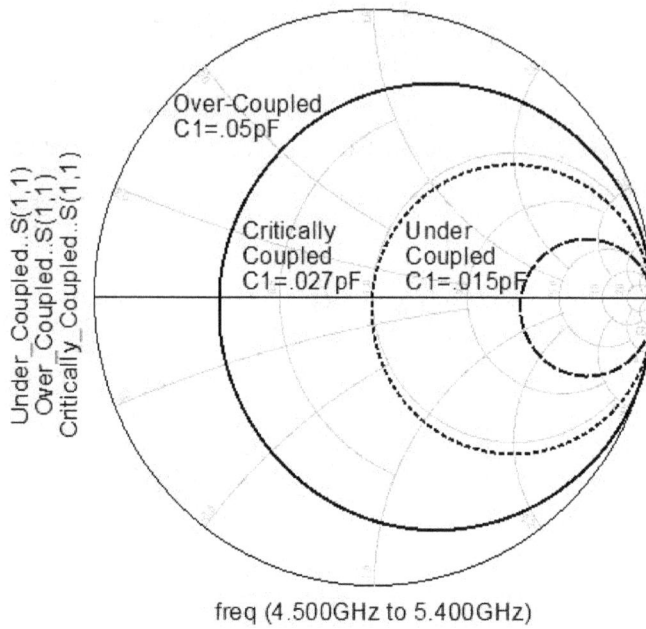

Figure 4-17 Smith Chart plots of the resonator coupling

Figure 4-18 shows the scalar plot of the input reflection coefficients.

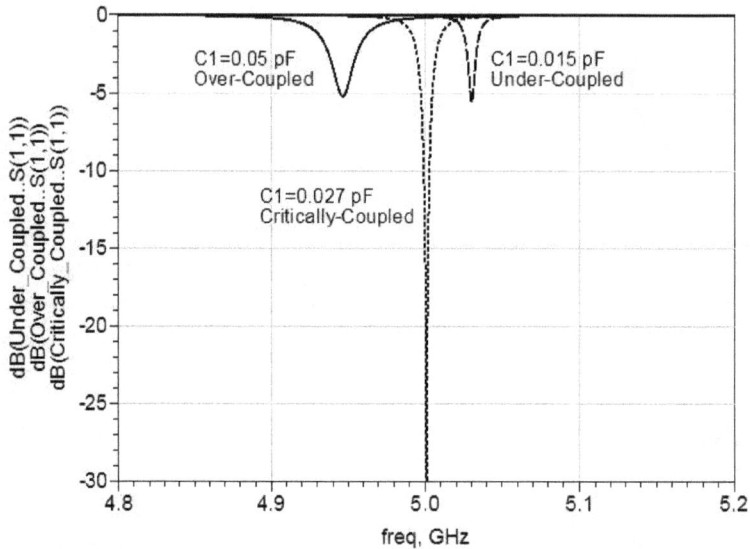

Figure 4-18 Scalar plots of the resonator input reflection coefficients

Filter Design at RF and Microwave Frequency

In Section 4.3 we have seen that it is possible to change the shape of the frequency response of a parallel resonant circuit by choosing different source and load impedance values. Likewise multiple resonators can be coupled to one another and to the source and load to achieve various frequency shaping responses. These frequency shaped networks are referred to as filters.

Lumped Element Filter Design

Classical filter design is based on extracting a prototype frequency-normalized model from a myriad of tables for every filter type and order. Fortunately these tables have been built into many filter synthesis software applications that are readily available. In this book we will examine the filter synthesis tool that is built into the ADS software. We will work through two practical filter examples, one low pass and one high pass using the ADS filter synthesis tool.

Low Pass Filter Design Example

Example 4.8-1A: Design a 145 MHz low pass filter for a satellite link system with the following requirements:

- Having a Chebyshev response with 0.1 dB pass band ripple
- Having a passband cutoff frequency (not -3 dB frequency) at 160 MHz
- Having at least -40 dB rejections at 435 MHz.

The transmitter and receiver antennas are on the same physical support boom so there is limited isolation between the transmitter and receiver. Even though the signals are at different frequencies, the broadband noise amplified by the power amplifier at 435 MHz will be received by the UHF antenna and sent to the sensitive receiver. Because the receiver is trying to detect very low signal levels, the received noise from the amplifier will interfere or 'de-sense' the received signals. Therefore it is necessary to design a 145 MHz Low Pass filter for this satellite link system. The specifications chosen for the filter design are selected as:

- Select a Chebyshev Response with 0.1 dB pass band ripple.
- Set the passband cutoff frequency (not the -3 dB frequency) at 160 MHz
- The reject requirement is at least -40 dB rejections at 435 MHz.
-

Solution: Create a new workspace and open a new schematic window. Use the Filter DesignGuide utility in ADS to design the lowpass filter. The procedure follows.

From the schematic window, click DesignGuide > Filter > Filter Control Window to open the Filter DesignGuide control window.

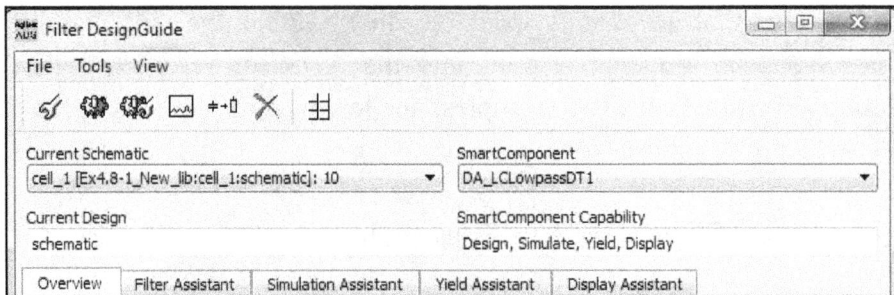

Figure 4-19 Filter DesignGuide control window

In the Filter DesignGuide control window, click View > Component Palette – All to place the filter SmartComponent Palette in the schematic window.

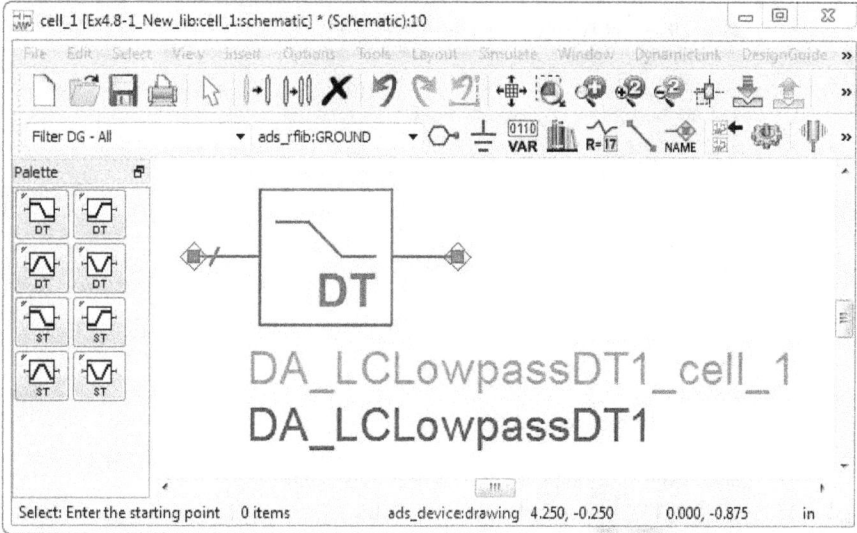

Figure 4-20 SmartComponent Palette in the schematic window

- From the list of Filter DG - All Palette in the schematic window, select the lowpass filter SmartComponenet, DA_LCLowpassDT.
- Click anywhere within the schematic window to place the component.
- Double-click the SmartComponent to open a dialog box containing all parameters.
- Modify the parameters to meet the design specifications, as shown in Figure 4-21

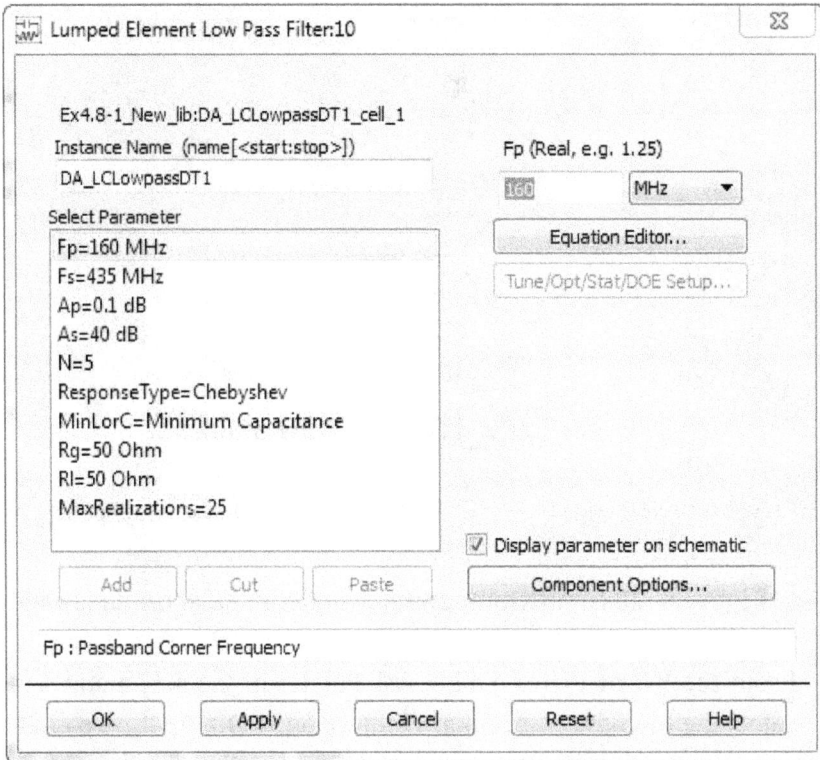

Figure 4-21 Modified parameters of the lowpass filter

- In the Filter DesignGuide control window select the Filter Assistant tab and click Design to start a simulation and generate the filter sub-network, as shown in Figure 4-22.

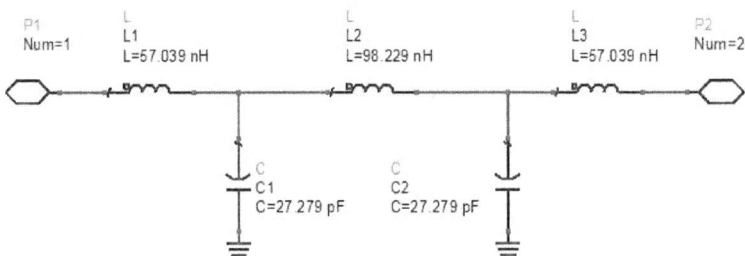

Figure 4-22 Lowpass filter sub-network (N=5)

The required filter order is determined by increasing the order until the specification of -40 dBc attenuation at 435 MHz is achieved. As the filter response curve shows, this Low Pass filter must be of fifth order to achieve the required attenuation.

After SmartComponents are designed and tested, they can be used as standalone components. The Filter DesignGuide is not needed to use them in new designs unless you want to modify or analyze them. In Figure 4-23 the lowpass filter SmartComponent is used in S parameter simulation circuit.

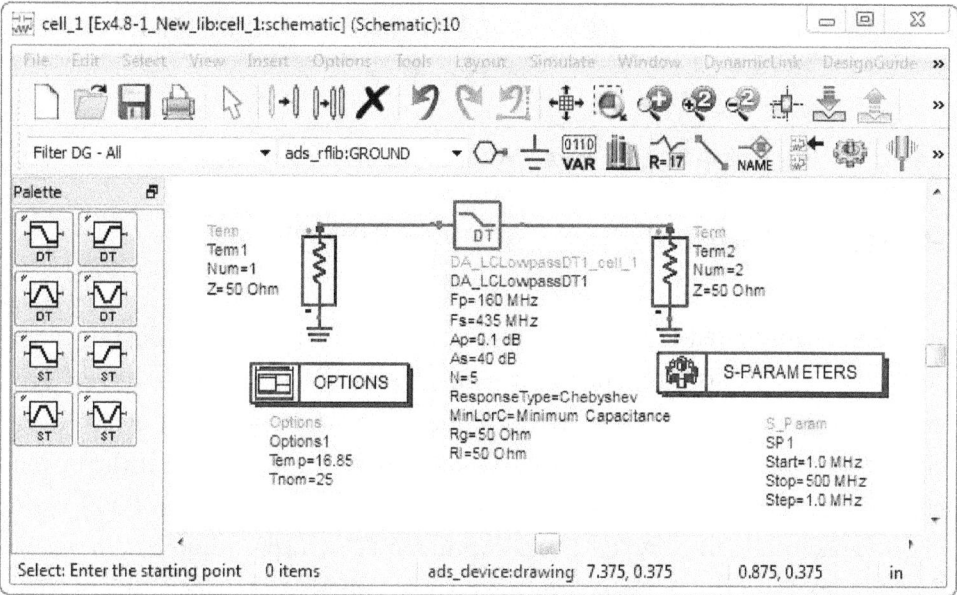

Figure 4-23 SmartComponent used in S parameter simulation

circuit Simulated Response of the lowpass filter is shown in Figure 4-24.

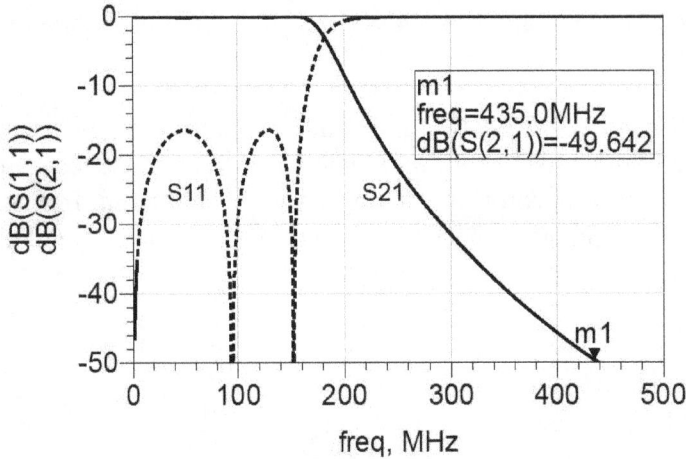

Figure 4-24 Simulated Response of the lowpass filter

Physical Model of the Low Pass Filter in ADS

It is important to realize that the synthesized filter is an ideal design in the sense that ideal (no parasitics and near infinite Q components have been used. To obtain a good 'real-world' simulation of the filter we need to use component models that have finite Q and parasitics such as multilayer chip capacitors for shunt capacitors. For the capacitors, choose the 6003 series chip capacitors from ATC Corporation. ATC Corporation has a useful application for selection of chip capacitors called ATC Tech Select. This program is available for free download from the ATC web site: www.atceramics.com. For the physical design of the filters the ATC chip capacitor models must be installed into the ADS software.

Because the filter is passing relatively high power (20 W, we cannot use small surface mount style chip inductors. Instead we will use air wound coils to realize the series inductors. The inductors will be realized with AWG#16 wire nickel-tin plated copper wire. The wire has a diameter of 0.05 inches or 50 mils. They will be wound on a 0.141 inch diameter form. Use the techniques covered in Chapter one, Section 1.4.1 to design the inductors using the Air Wound inductor model in ADS. The filter model is then reconstructed using the physical inductor models for the series inductors. Make sure to model the substrate and the interconnecting printed circuit board traces as microstrip lines. Also model the ground connection of the shunt capacitors as a microstrip via hole. Although these PCB parasitic effects are normally more pronounced at frequencies above 2 GHz, it is often surprising the effect that these parasitics have at lower frequencies.

Example 4.8-1B: The physical filter model is shown in Figure 4-25.

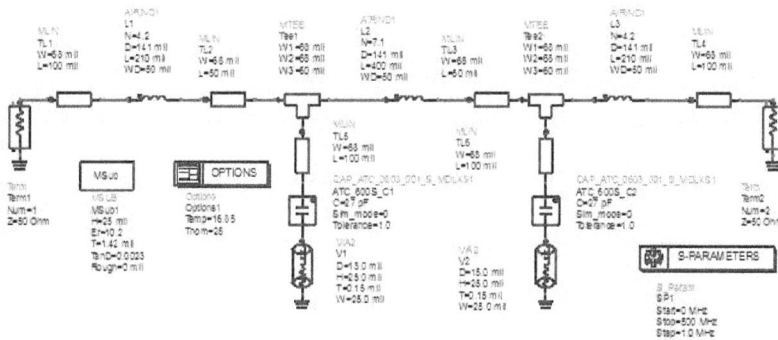

Figure 4-25 Simulated Response of the lowpass

filter The physical filter response is shown in Figure 4-26.

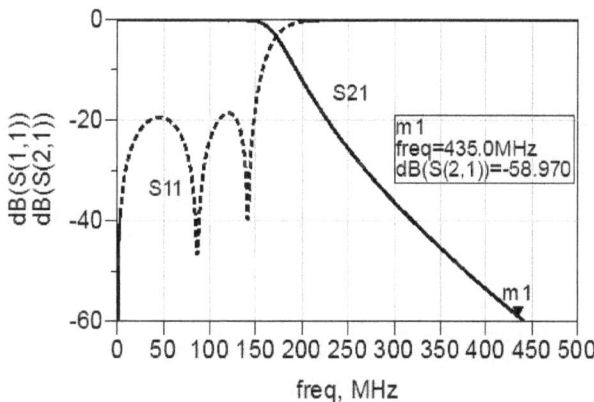

Figure 4-26 Simulated Response of the final lowpass filter

The response shows that the attenuation specification has been achieved.

High Pass Filter Design Example

Example 4.8-2: Design a high pass filter that passes frequencies in the 420 MHz to 450 MHz range. This filter could be placed in front of the preamp used in the downlink of the satellite system. This would help to keep out any of the transmit energy or noise power in the 146 MHz transmit frequency range. The High Pass Filter specifications are:

- The pass band cutoff frequency (not the -3 dB frequency) is 420 MHz.
- The filter has a Chebyshev response with 0.1 dB pass band ripple. The reject requirement is at least -60 dB rejections at 146 MHz.

Solution: Create a new workspace in ADS and open a new schematic window. From the schematic window, click DesignGuide > Filter > Filter Control Window to open the Filter DesignGuide control window. In the Filter DesignGuide control window, click View > Component Palette – All to place the filter SmartComponent Palette in the schematic window. From the list of Filter DG - All Palette select the highpass filter SmartComponenet, DA_LCHighpassDT, and click anywhere within the schematic window to place the component. Modify the component parameters to meet the highpass filter design specifications, as shown in Figure 4-27.

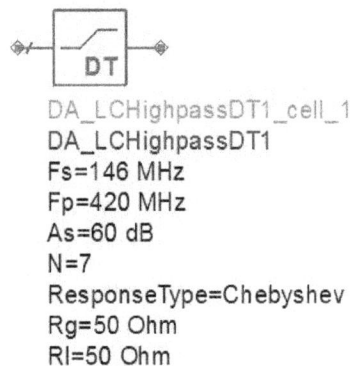

DA_LCHighpassDT1_cell_1
DA_LCHighpassDT1
Fs=146 MHz
Fp=420 MHz
As=60 dB
N=7
ResponseType=Chebyshev
Rg=50 Ohm
Rl=50 Ohm

Figure 4-27 Parameters of the highpass filter

The required filter order is determined by increasing N until the specification of -60 dBc attenuation at 146 MHz is achieved. It is always a good practice to design for some additional rejection (margin) that exceeds the minimum requirement, as shown in Figure 4-28. The Filter DesignGuide session recommends a filter of 7th order.

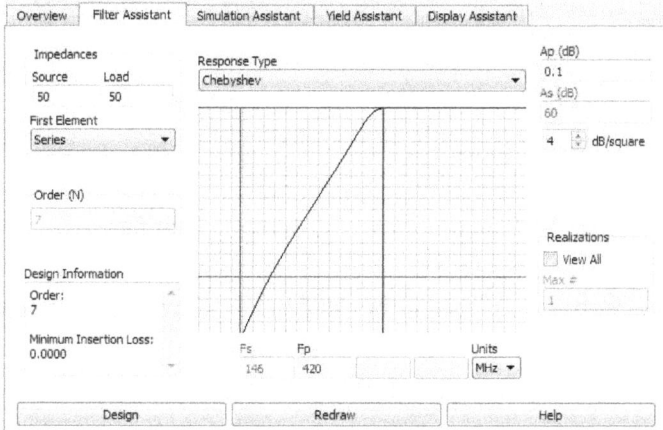

Figure 4-28 DesignGuide highpass filter response

The highpass filter sub-network is shown in Figure 4-29.

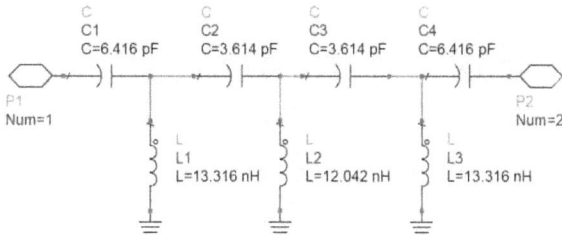

Figure 4-29 Lumped element model of the high pass filter (N=7)

Attach 50 Ohm terminations to the highpass filter SmartComponent model and simulate the schematic from 100 to 500 MHz, as shown in Figure 4-31.

Term
Term1
Num=1
Z=50 Ohm

DT

DA_LCHighpassDT1_cell_1
DA_LCHighpassDT1
Fs=146 MHz
Fp=420 MHz
As=60 dB
N=7
ResponseType=Chebyshev
Rg=50 Ohm
Rl=50 Ohm

Term
Term2
Num=2
Z=50 Ohm

OPTIONS

Options
Options1
Temp=16.85
Tnom=25

S-PARAMETERS

S_Param
SP1
Start=100 MHz
Stop=500 MHz
Step=1 MHz

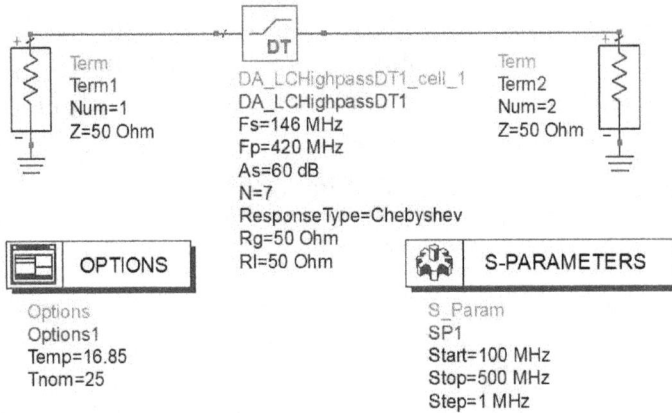

Figure 4-30 Simulation of the high pass filter

The simulated response of the highpass filter is shown in Figure 4-31.

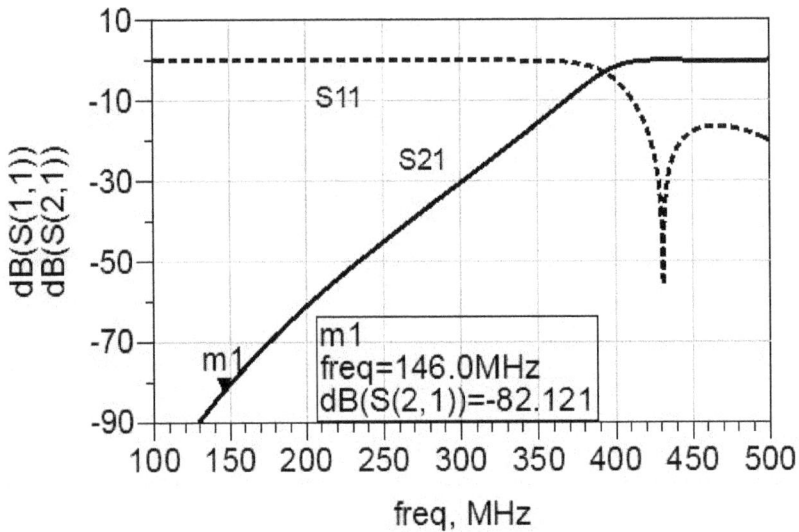

m1
freq=146.0MHz
dB(S(2,1))=-82.121

Figure 4-31 Response of the high pass filter

As Figure 4-31 shows the attenuation of -60 dBc at 146 MHz has been achieved with a good margin.

Microstrip Stepped Impedance Low Pass Filter Design

In the microwave frequency region filters can be designed using distributed transmission lines. Series inductors and shunt capacitors can be realized with microstrip transmission lines. In the next section we will explore the conversion of a lumped element low pass filter to a design that is realized entirely in microstrip.

Example 4.9-1A: Design a lumped element 2.2 GHz Chebyshev-response lowpass filter having 0.1 dB passband ripple and -40 dB rejection at 6 GHz.

Solution: Following the procedure discussed in Example 4.8-1, the SmartComponent design of the lowpass filter is show in Figure 4-32.

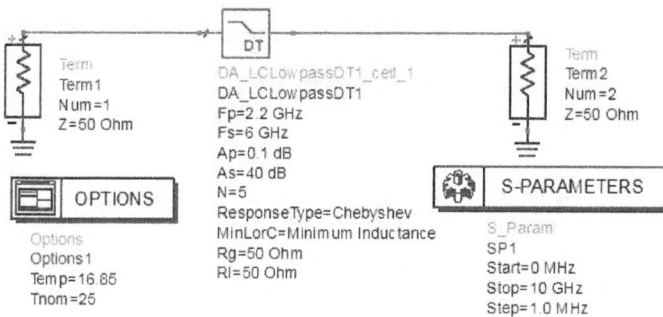

Figure 4-32 Lumped element design of the 2.2 GHz low pass filter

The lumped element design of the filter is shown in Figure 4-33.

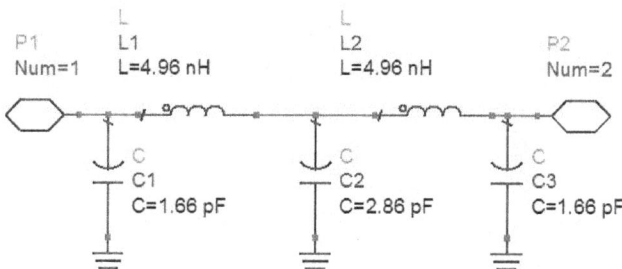

Figure 4-33 Lumped element 2.2 GHz low pass filter

The simulated response of the filter is shown in Figure 4-34.

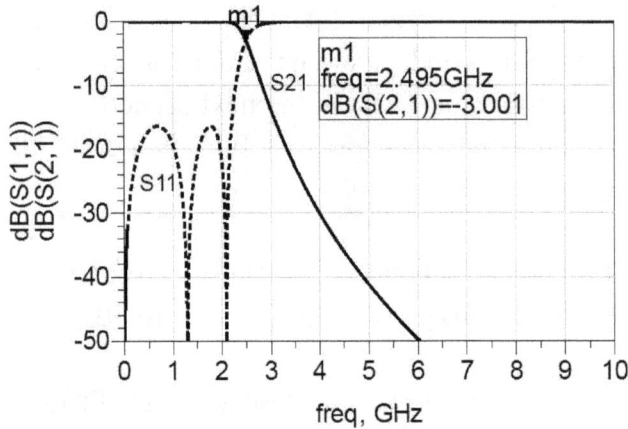

Figure 4-34 Response of the lumped element 2.2 GHz lowpass filter

The marker m1 in Figure 4-46 shows that the 3 dB bandwidth of the lowpass filter is 2495 MHz.

Lumped Element to Distributed Element Conversion

Example 4.9-1B: Convert all the lumped elements of the filter in Example 4.9-1A to microstrip transmission lines. The microstrip substrate is Rogers's 6010 material (ε_r= 10.2)with a 0.025 inch dielectric thickness.

Solution: After a Filter DesignGuide SmartComponent has been designed, the lumped inductors and capacitors can be transformed into equivalent distributed element counterparts using the ADS Transformation Assistant. The Filter Transformation Assistant helps us to quickly and easily transform an ideal filter topology to a form that is realizable for RF and microwave systems.

The Transformation Assistant is opened from the Filter DesignGuide Control window, by selecting When the Transformation Assistant is opened, the SmartComponent subnetwork appears in the schematic window and a dialog box is opened. The transformations are then accomplished using the controls on the dialog.box.

After the Transformation Assistant has been selected, the graphical area displays the components that can be transformed. Black components represent elements included in the original circuit available for transformation, while gray components represent elements not included in the original circuit.

Check the LC, TLine to Microstrip transformations and select the series inductors L1 and L2 to be transformed to microstrip transmission lines. Set the characteristic impedance of the microstrip line to $Z_0 = 80$ Ohm and enter the microstrip substrate parameters, as shown in Figure 4-35.

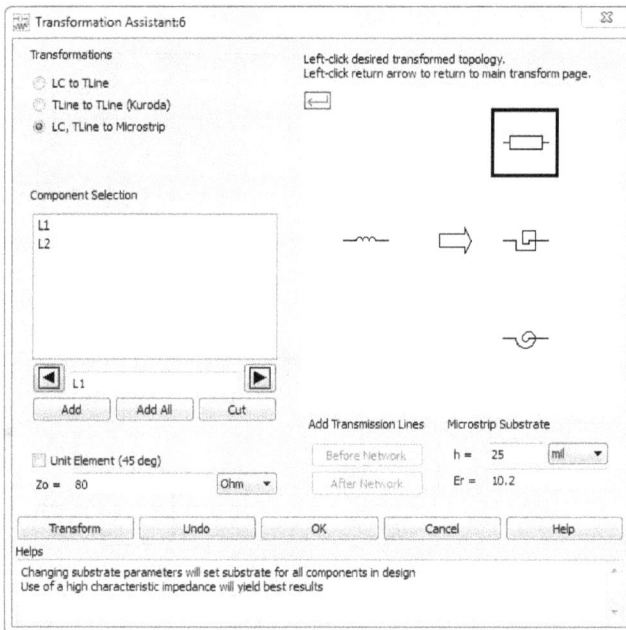

Figure 4-35 Transformation of series inductors to 80 Ω microstrip line

Click Transform and then click OK. As Figure 4-36 shows the series inductors L1 and L2 have been transformed to 80 Ω microstrip transmission lines on the microstrip substrate.

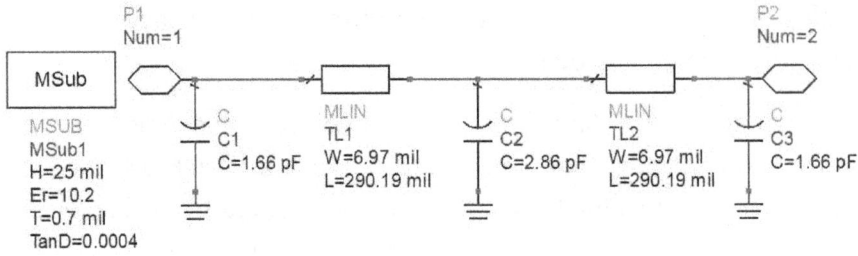

Figure 4-36 Series inductors transformed to microstrip lines

From the graphical area, use the left red arrow to select the shunt capacitors for transformation. The graphical area changes to reveal the other distributed element equivalents available for substitution. Repeat the same process for the capacitors but select $Z_0 = 20$ Ohm, as shown in Figure 4-37.

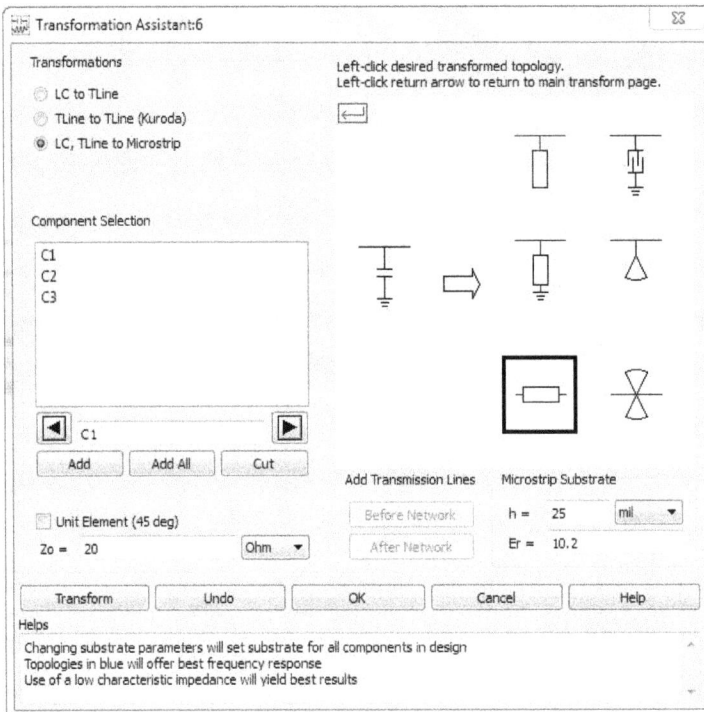

Figure 4-37 Transforming the shunt capacitors to 20 Ω microstrip line

Click Transform and then click OK. Now the capacitors have also been transformed to microstrip lines. Adding a short 50 Ω section to the input and output, the initial schematic of the distributed low pass filter, as shown in Figure 4-38, has been generated.

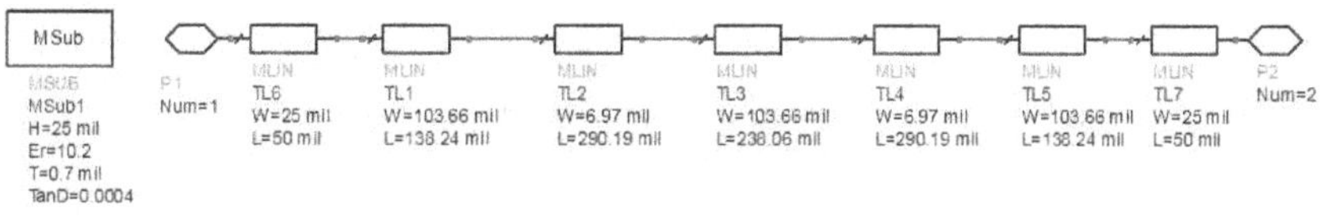

Figure 4-38 Initial schematic of the distributed low pass filter

Simulate the distributed lowpass filter and display the response from 0 to 10 GHz.

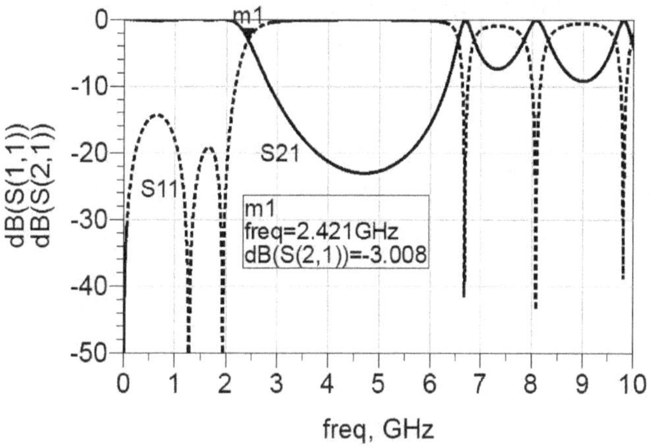

Figure 4-39 Initial response of the low pass filter

Note that the marker m1 in Figure 4-39 shows that the 3 dB bandwidth of the lowpass filter has slightly decreased to 2421 MHz.

The printed circuit board layout of the initial lowpass filter is shown in Figure 4-40.

PCB Layout

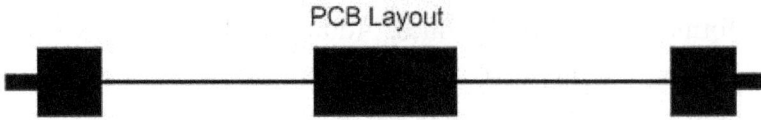

Figure 4-40 Initial PCB layout of the lowpass filter

Examine the printed circuit board, PCB, layout of the lowpass filter of Figure 4-40. Note the change in geometry as the impedance transitions from 50 Ω at the input to 20 ‚Ω representing the shunt capacitor, and then from 20 Ω to 80 ‚Ω representing the series inductor. These abrupt changes in geometry are known as discontinuities. Discontinuities in geometry result in fringing capacitance and parasitic inductance that will modify the frequency response of the circuit. At RF and lower microwave frequencies (up to about 2 GHz) the effects of discontinuities are minimal and sometimes neglected [4]. As the operation frequency increases, the effects of discontinuities can significantly alter the performance of a microstrip circuit. ADS software has several model elements that can help to account for the effects of discontinuities. These include: T-junctions, cross junctions, open circuit end effects, coupling gaps, and bends.

Example 4.9-1C: A Microstrip Step element can be placed between series lines of abruptly changing geometry to account for the step discontinuity. Place the Microstrip Step element at each impedance transition in the filter. Make sure that the narrow side and wide side are directed appropriately.

Solution: The Microstrip Step element, MSTEP, will automatically use the adjacent width in its calculation. Figure 4-41 shows the lowpass stepped impedance filter with the step elements added between each transmission line section.

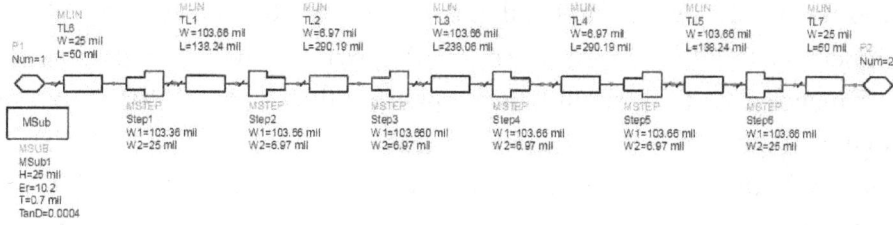

Figure 4-41 Stepped impedance filter with added "step" elements

The simulated response of the modified filter is shown in Figure 4-42.

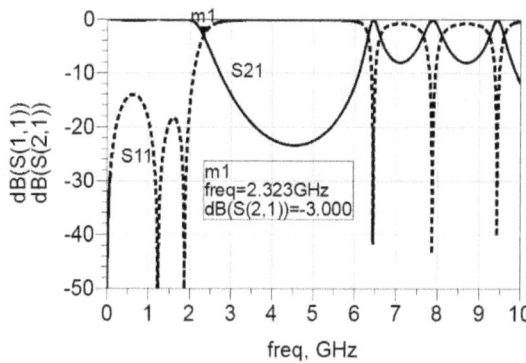

Figure 4-42 Filter frequency shift due to step discontinuities

A comparison of the initial lowpass filter model response in Figure 4-39 and the modified model response in Figure 4-42 shows that there is some frequency shift in the filter response as it is evidenced, about 100 MHz, by the 3 dB bandwidth.

Microstrip Coupled Line Filter Design

The edge coupled microstrip line is very popular in the design of bandpass filters. A cascade of half-wave resonators in which quarter wave sections are parallel edge coupled lines, are very useful for realizing narrow band, band pass filters. This type of filter can typically achieve $\leq 15\%$ fractional bandwidths.

Example 4.9-2: Design a band pass filter at 10.5 GHz. The filter is designed on RO3010 substrate (ε_r = 10.2) with a dielectric thickness of 0.025 inches. The filter should have a pass band of 9.98 – 11.03 GHz. As a design goal the filter should achieve at least 20 dB rejection at 9.65 GHz. In other words the filter is required to have > 20 dB rejection at 330 MHz below the lower passband frequency.

Solution: The Passive Circuit DesignGuide utility in ADS is used to design the filter network. The procedure for opening and using the Passive Circuit DesignGuide utility is as follows.

1. In the schematic window, choose DesignGuide > Passive Circuit > Passive Control Window.

2. From Passive Circuit DesignGuide toolbar, choose View > Component Pallet > Microstrip.

3. From Passive Circuit DG – Microstrip pull-down list, choose Filter-Bandpass

4. From Filter-Bandpass pull-down list, choose Chebyshev

5. Set the filter parameters and vary the order until the desired filter rejection is achieved.

The Passive Circuit DesignGuide program shows that a 6th order filter should meet the rejection specifications. The synthesized filter schematic and response are shown in Figures 4-43 and 4-44.

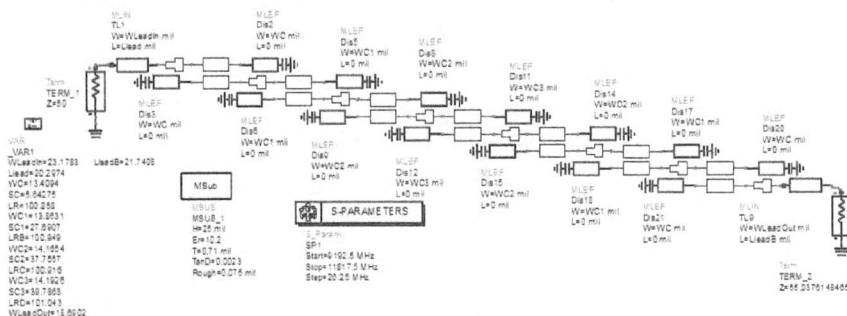

Figure 4-43 Parallel line edge coupled microstrip bandpass filter schematic

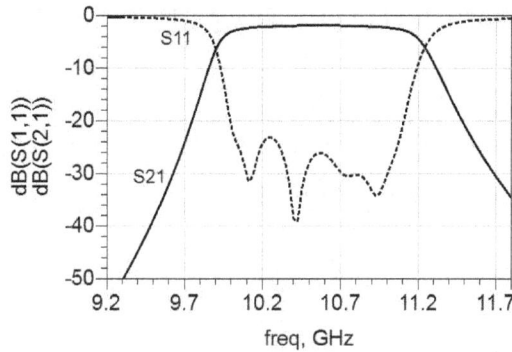

Figure 4-44 Synthesized bandpass filter response

References and Further Reading

[1]   Ali A. Behagi, *RF and Microwave Circuit Design,* A Design Approach Using (**ADS**). Techno Search, Ladera Ranch, CA 92694. August 2015.

[2]   RF Circuit Design, Second Edition, Christopher Bowick, Elsevier 2008

[3]   Keysight Technologies, Manuals for Advanced Design System, *ADS 2016.01 Documentation Set*, Keysight EEsof EDA Division, Santa Rosa, California, www.keysight.com

[4]   Foundations for Microstrip Circuit Design, T.C. Edwards, John Wiley & Sons, New York, 1981

[5]   Q Factor, Darko Kajfez, Vector Fields, Oxford Mississippi, 1994

[6]   David M. Pozar, *Microwave Engineering*, Second Edition, John Wiley and Sons, Inc. 1998

[7]   Q Factor Measurement with Network Analyzer, Darko Kajfez and Eugene Hwan, IEEE Transactions on Microwave Theory and Techniques, Vol. MTT-32, No. 7, July 1984.

[8]   High Frequency Techniques, Joseph F. White, John Wiley & Sons, Inc., 2005

[9]   Soft Substrates Conquer Hard Designs, James D. Woermbke, Microwaves, January 1982.

[10]   Principles of Microstrip Design, Alam Tam, RF Design, June 1988.

[11]   David M. Pozar, *Microwave Engineering*, Third Edition, John Wiley and Sons, 2005.

Problems

4-1.   Consider the one port resonator that is represented as a series RLC circuit as shown. Analyze the circuit, with R = 5 Ω, L = 5 nH, and C = 5 pF. Plot the magnitude of the resonator input impedance and measure the resonance frequency.

4-2.   Consider the one port resonator that is represented as a parallel RLC circuit as shown. Analyze the circuit, with R = 500 Ω, L = 50 nH, and C = 50 pF. Plot the magnitude of the resonator input impedance and measure the resonance frequency.

4-3.   Design a Butterworth lowpass filter having a passband of 2 GHz with an attenuation 20 dB at 4 GHz. Plot the insertion loss versus frequency from 0 to 5 GHz. The system impedance is 50 Ω.

4-4.   Design a 5th order Chebyshev highpass filter having 0.2 dB equal ripples in the passband and cutoff frequency of 2 GHz. The system impedance is 75 Ω. Plot the insertion loss versus frequency from 0 to 5 GHz.

4-5.   In a full duplex communication link, the uplink signal is around 200 MHz while the downlink is at 500 MHz. A 25 Watt power amplifier is used on the uplink with 20 dB gain.

(a) Design a low pass filter on the uplink to pass the 200 MHz uplink signal while rejecting any noise power in the 500 MHz band. Design the passband cutoff frequency is at 220 MHz, therefore, the filter should have a Chebyshev Response with 0.1 dB pass band ripple. The reject requirement is at least -40 dB rejections at 500 MHz.

(b) Design a High Pass Filter that passes frequencies in the 480 MHz to 520 MHz range. The High Pass Filter specifications are: The passband cutoff frequency is 480 MHz, therefore, the filter should have a Chebyshev response with 0.1 dB pass band ripple. The reject requirement is at least -60 dB losses at 190 MHz.

4-6.   Design a 75 Ω transmission line of sufficient length to act as a 10 nH inductor. Calculate the microstrip line width on the Rogers 0.025 inch RO3010 material.

4-7.   Using the microwave filter synthesis tool, design a stepped impedance low pass filter on RO3003 material that is 0.010 inches thick. Use a Chebyshev response with a 0.01dB ripple and a cutoff frequency of 4 GHz. Determine the worst case in band return loss and the rejection at 6 GHz.

4-8.   For the filter design of Problem 4-7 create an EM simulation using Momentum. Compare the EM simulation to a linear simulation. Comment on the rejection comparison at 6 GHz.

4-9.   For the filter design of Problem 4-7 determine the frequencies at which reentrant modes exist up through 20 GHz.

4-10.   Design a half wave microstrip resonator at 10 GHz using RO3003 substrate that is 0.020 thick. Initially design the resonator with a 50 W line impedance. Select a coupling capacitor to critically couple the resonator to the 50 Ω source. Then determine the resonator line impedance that results in the highest unloaded Q_o.

4-11.   Use the microwave filter synthesis tool to design a parallel edge
        coupled filter on RO3003 substrate that is 0.010 thick. Use a
        Chebyshev characteristic with 0.10 dB ripple. Design the passband
        to cover 10.7 GHz to 12.2 GHz. Determine the filter order required
        to achieve 30 dB rejection at 9 GHz.

4-12.   For the filter design of Problem 4-10, create an EM simulation using
        Momentum. Compare the linear and EM simulations. Determine the
        minimum box width (EM Box height, Y) that creates a box
        resonance frequency. What is the frequency of the box resonance.

Chapter 5

Power Transfer and Impedance Matching

Introduction

Impedance matching is an integral part of RF and microwave circuit and system design. It is necessary for the efficient transfer of power from a source to the load. For example, in microwave amplifier design, the need for impedance matching arises when the amplifier must be properly terminated at both terminals in order to deliver maximum power from the source to the load. In narrowband applications impedance matching can be achieved, at a single frequency, with a lossless two-element network, known as L-network. In this chapter the basics of power transfer and the conditions for maximum power transfer are presented. The mathematical equations for the design of discrete L-networks are derived using Matlab compatible expressions.

Maximum Power Transfer Conditions

For the network of Figure 5-1 a voltage source, V_S, and the series impedance $Z_S = R_S + jX_S$ are connected to a network, having the input impedance $Z_{IN} = R_{IN} + jX_{IN}$, the power transferred to the network is given by:

$$P_{Network} = \frac{\text{Re}\left[V_{IN} I_{IN}^*\right]}{2}$$

In this equation Re denotes the real part and the symbol * denotes the conjugate value.

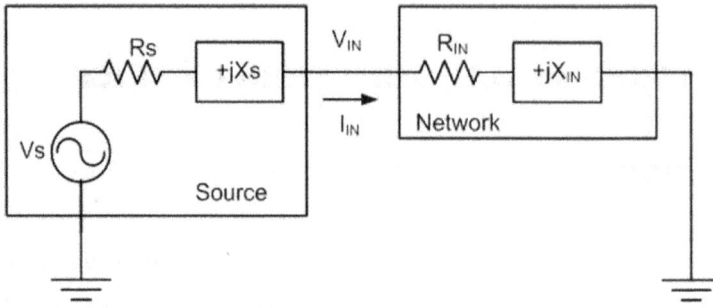

Figure 5-1 Voltage source connected to complex load impedance

The condition for maximum power transfer is that the load impedance be equal to the conjugate of the source impedance given by:

$$Z_{IN} = Z_S^*$$

For maximum power transfer, two cases are considered.

Case 1. When Source and Load Are Real Impedances

Figure 5-2 shows the case when source and load impedances are real.

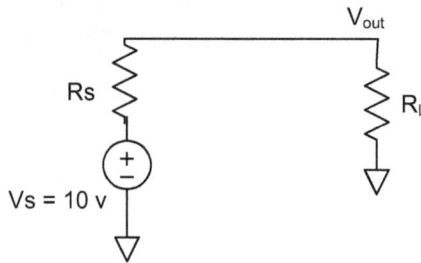

Figure 5-2 Network with purely resistive source and load impedance

Example 5.2-1: Use the ADS linear simulation to prove the maximum power transfer condition in a purely resistive system.

Solution: This is a very simple ADS schematic with only input and output ports, as shown in Figure 5-3. The input port represents an RF signal source with 50 Ω source impedance and the output port is simply a 50 Ω resistive load. Simulate the schematic from 450 to 550 MHz and add a rectangular graph with the insertion loss, S21, in dB. Figure 5-3 shows the schematic and plot indicating that S21 = 0 dB. Because there is zero insertion loss between the source and load, there is maximum power transfer.

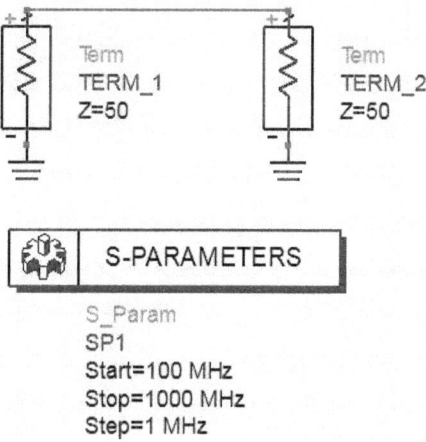

Term
TERM_1
Z=50

Term
TERM_2
Z=50

S-PARAMETERS

S_Param
SP1
Start=100 MHz
Stop=1000 MHz
Step=1 MHz

Figure 5-3 Case I power transfer with $R_L = R_S$

Simulate the schematic and display the insertion loss, S21, in a rectangular plot as shown in Figure 5-4.

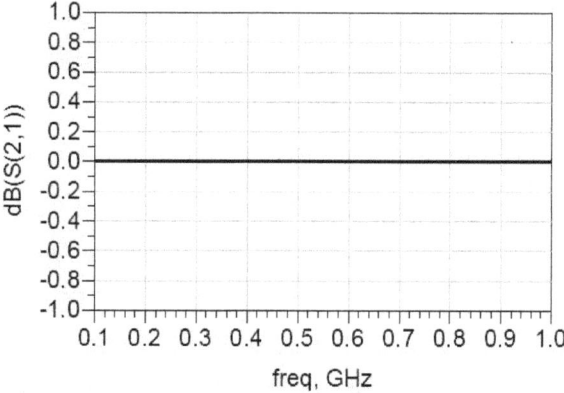

Figure 5-4 Insertion Loss with $R_L = R_S$

Example 5.2-2: Use the ADS linear simulation to prove the maximum power transfer condition for a load resistance of~ Z = 25 Ohm.

Solution: Change the load impedance to 25 Ω as shown in Figure 5-5.

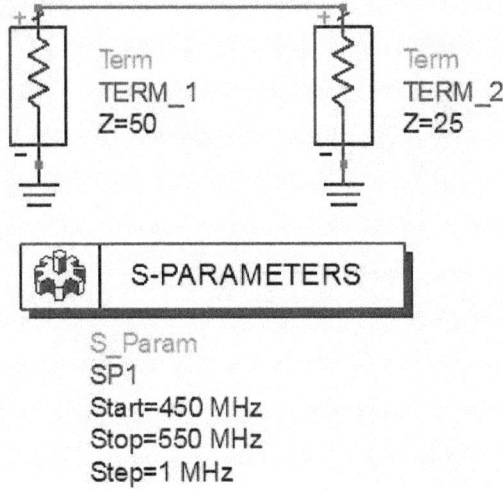

Figure 5-5 Case II power transfer with $R_L < R_S$

Simulate the schematic and display the insertion loss, S21, in a rectangular plot as shown in Figure 5-6.

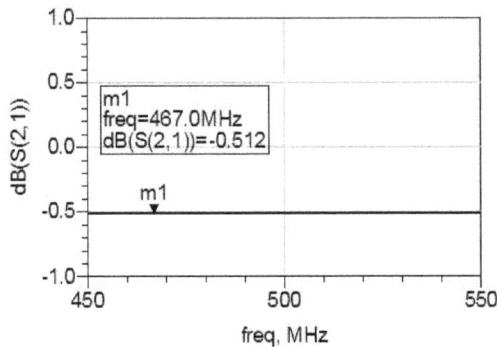

Figure 5-6 Insertion Loss with $R_L < R_S$

Notice that S21 = -0.512 dB indicating insertion loss. To calculate the amount of power loss, we know that the maximum power transfer is 0.5

Watts. To determine the amount of 0.512 dB power loss we first convert the 0.5 Watts to dBm which is the power in dB relative to 1 mW. Utilizing the conversion equation,

$$10 \log (\text{power in mW}) = \text{power in dBm}.$$

$$10 \log (500mW) = 26.98 \ dBm$$

The loss of 0.512 dB is subtracted from the 26.98 dBm to result in (26.98 dBm − 0.512 dB) = 26.47 dBm. Then converting from dBm back to mW we get 444 mW as calculated in Example 5.2-1.

$$10^{\left(\frac{26.47}{10}\right)} = 444 \ mW \ \ or \ \ 0.444 \ Watts$$

This proves that the maximum power is transferred when $R_S = R_L$.

Case 2. When Load is a Complex Impedance

We know that maximum power transfer occurs when $Z_S = Z_L*$. Therefore, if $Z_L = R_L - jX_L$, then for maximum power transfer we must have $Z_S = R_L + jX_L$.

Example 5.2-3: If the load is 50 Ω in series with a 15 pF series capacitance, find the source impedance to have maximum power transfer at 500 MHz.

Solution: We can find a source inductance that cancels the reactance of the load at this frequency. The reactance of the 15 pF capacitor is:

$$X_C = \frac{1}{2 \pi f C} = 21.231 \ \Omega$$

Therefore,

$$Z_L = 50 - j21.231 \ \Omega$$

For maximum power transfer the source impedance must be:

$$X_S = 50 + j21.231 \quad \Omega$$

At 500 MHz the value of the source series inductor is:

$$L = \frac{21.231}{2\pi f} = 6.76 \quad nH$$

To demonstrate the maximum power transfer, create a schematic in ADS as shown in Figure 5-7. Place the specified 15 pF capacitance in series with the load and the 6.76 nH inductance in series with the source. Plot the insertion loss and VSWR on a rectangular graph as shown in Figure 5-8. Note that unlike the purely resistive source and load case, a complex conjugate match occurs at a single frequency. A perfect 1:1 VSWR is achieved at the conjugate match frequency of 500 MHz.

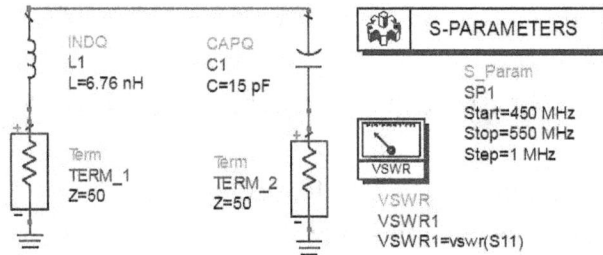

Figure 5-7 Maximum power transfer with complex impedance

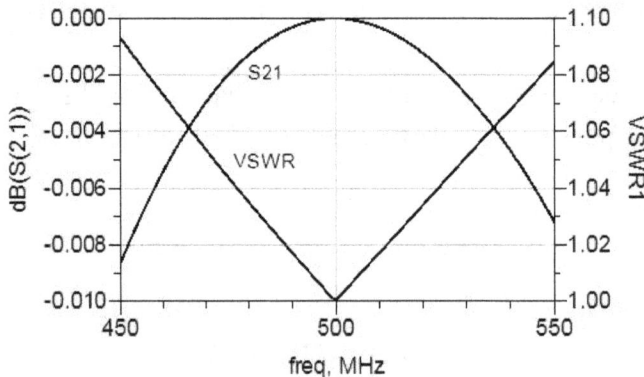

Figure 5-8 VSWR and insertion loss with complex source and load

Analytical Design of Impedance Matching Networks

One of the important tasks in RF and microwave engineering is the determination of how an arbitrary complex load impedance, $Z_L = R_L + jX_L$, is analytically matched to any complex source impedance, $Z_S = R_S + jX_S$, as shown in Figure 5-9. This problem arises mainly in the design of inter-stage matching networks between active devices or between an antenna and a transmitter.

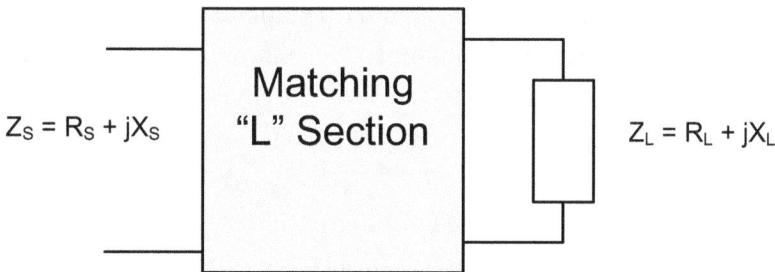

$Z_S = R_S + jX_S$ Matching "L" Section $Z_L = R_L + jX_L$

Figure 5-9 General impedance matching with an L-network

Example 5.3-1A: Design L-networks matching a complex load impedance, $Z_L = 10 - j15\ \Omega$, to a complex source impedance, $Z_S = 15 - j20\ \Omega$, at 2 GHz. This problem has four solutions. The solutions are given separately.

First Solution: We utilize Equations (5-7) and (5-8) in the Appendix B to calculate the element values of the matching L-network. The procedure follows.

1. Enter design parameters

RS = 15; XS = -20; RL = 10; XL = -15; f = 2e9

2. Use Equations (5-7) and (5-8) to calculate the matching element values

B1=((RL*XS)+sqrt(RS*RL*(RS^2+XS^2-RS*RL)))/(RL*(RS^2+XS^2))

X1=(RL*XS-RS*XL)/RS+(RS-RL)/(B1*RS)

L1=X1/(2*pi*f)

C2=B1/(2*pi*f)

The calculation results show that B1 = 0.011 and X1 = 32.795, therefore, the series element is an inductor, L1 = 2.61 nH, and the shunt element is a capacitor, C2 = 0.852 pF.

To verify the solution, create a new workspace in ADS and open a new schematic window. Insert the S_Params Template and add capacitor and inductor from the Lumped-Components palette. Connect the matching elements as shown in Figure 5-10.

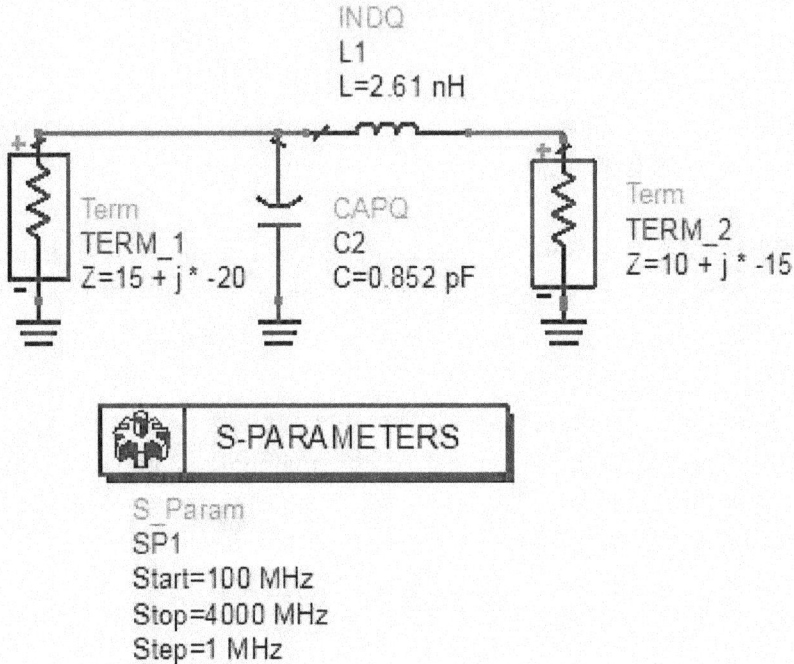

INDQ
L1
L=2.61 nH

Term
TERM_1
Z=15 + j * -20

CAPQ
C2
C=0.852 pF

Term
TERM_2
Z=10 + j * -15

S-PARAMETERS

S_Param
SP1
Start=100 MHz
Stop=4000 MHz
Step=1 MHz

Figure 5-10 Simulation of the first matching L-network

Simulate the schematic and display S11 and S21 in a rectangular plot.

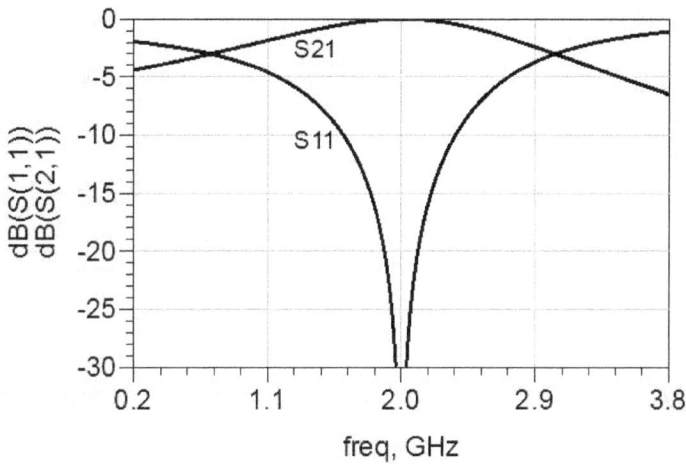

Figure 5-11 Response of the first matching L-network

The simulated response in Figure 5-11 shows that the matching L-network perfectly matches the load impedance to the source impedance at 2 GHz.

Example 5.3-1B: Design the second L-network to match the complex load impedance, $Z_L = 10 - j15\ \Omega$, to a complex source impedance, $Z_S = 15 - j20$ Ω, at 2 GHz.

Second Solution: Use Equations (5-9) and (5-10) in the Appendix B to calculate the second solution. The procedure follows.

1. Enter Design Parameters

RS=15; XS=-20; RL=10; XL=-15; f=2e9

2. Use Equations (5-9) and (5-10) in the textbook to calculate the matching element values

B2=((RL*XS)-sqrt(RS*RL*(RS^2+XS^2-RS*RL)))/(RL*(RS^2+XS^2))

X2=(RL*XS-RS*XL)/RS+(RS-RL)/(B2*RS)

C1=-1/(2*pi*f*X2)

L2=-1/(2*pi*f*B2)

The calculation results show that B2= -0.075 and X2 = -2.795, therefore, the series element is a capacitor, C1 = 28.47 pF and the shunt element is an inductor, L2 = 1.065 nH.

To plot the response of the matching network, create a new workspace in ADS and open a new schematic window. Insert the S_Params Template and add capacitor and inductor from the Lumped-Components palette. Connect the matching elements as shown in Figure 5-12.

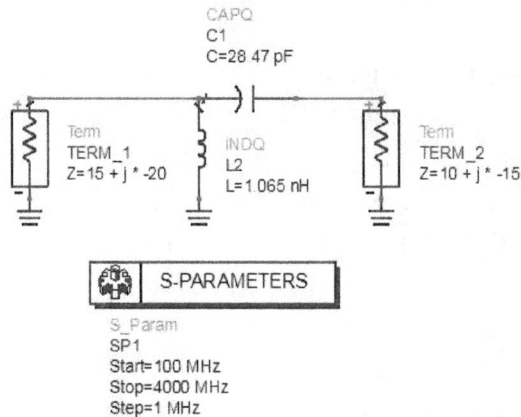

Figure 5-12 Simulation of the second matching L-network

Simulate the schematic from 0 to 4000 MHz and display the return loss, S11, and insertion loss, S21, in dB in a rectangular plot, as shown in Figure 5-13.

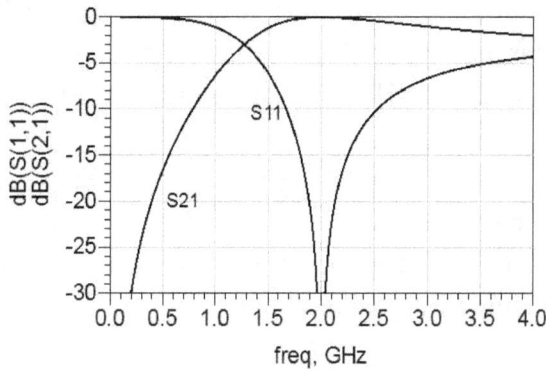

Figure 5-13 Response of the second matching L-network

The simulated response in Figure 5-13 shows that the matching L-network perfectly matches the load impedance to the source impedance at 2 GHz.

Example 5.3-1C: Design the third L-network to match the complex load impedance, $Z_L = 10 - j15 \ \Omega$, to a complex source impedance, $Z_S = 15 - j20 \ \Omega$, at 2 GHz.

Third Solution: Use Equations (5-12) and (5-13) in the Appendix B to calculate $B3$ and $X3$ for the third solution. The procedure follows.

1. Enter design parameters

RS=15; XS=-20; RL=10; XL=-15; f=2e9

2. Use Equations (5-12) and (5-13) in the textbook to calculate the matching element values

B3=((RS*XL)+sqrt(RS*RL*(RL^2+XL^2-RS*RL)))/(RS*(RL^2+XL^2))

X3=(RS*XL-RL*XS)/RL+(RL-RS)/(B3*RL)

L1=-1/(2*pi*f*B3)

L2=X3/(2*pi*f)

The calculation results show that B3 = -0.013 and X3 = 36.202, therefore, the shunt element is an inductor L1= 6.16 nH and the series element is another inductor L2=2.881 nH, as shown in Figure 5-14.

To plot the response of the matching network, create a new workspace in ADS and open a new schematic window. Insert the S_Params Template and add capacitor and inductor from the Lumped-Components palette. Connect the matching elements as shown in Figure 5-14.

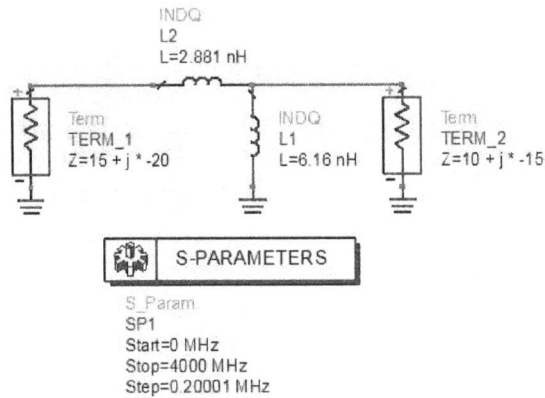

Figure 5-14 Simulation of the third matching L-network

Simulate the schematic and display S11 and S21 in a rectangular plot.

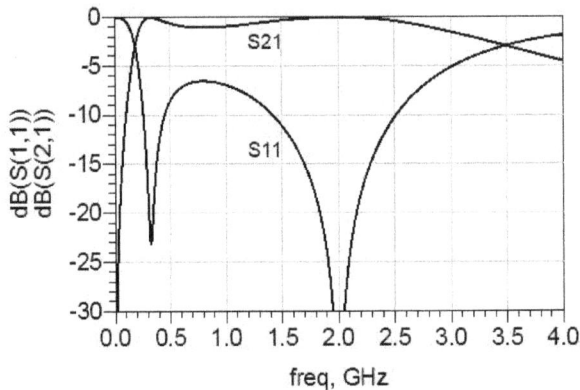

Figure 5-15 Response of the third matching L-network

The simulated response in Figure 5-15 shows that the matching L-network perfectly matches the load impedance to the source impedance at 2 GHz.

Example 5.3-1D: Design the fourth L-network to match the complex load impedance, $Z_L = 10 - j15 \ \Omega$, to a complex source impedance, $Z_S = 15 - j20 \ \Omega$, at 2 GHz.

Fourth Solution: Use Equations (5-14) and (5-15) in the Appendix B to calculate *B4* and *X4*.

1. Enter design parameters:

RS=15; XS=-20; RL=10; XL=-15; f=2e9

2. Use Equations (5-14) and (5-15) in Matlab script to calculate the matching element values

B4=((RS*XL)-sqrt(RS*RL*(RL^2+XL^2-RS*RL)))/(RS*(RL^2+XL^2))

X4=(RS*XL-RL*XS)/RL+(RL-RS)/(B4*RL)

L1=-1/(2*pi*f*B4)

L2=X4/(2*pi*f)

The calculation results show that B4 = -0.079 and X4 = 3.798, therefore, the shunt element is an inductor L1=1.002 nH and the series element is another inductor L2=0.302 nH.

To plot the response of the matching L-network, create a new workspace in ADS and open a new schematic window. Insert the S_Params Template and add capacitor and inductor from the Lumped-Components palette. Connect the matching elements as shown in Figure 5-16.

INDQ
L2
L=0.302 nH

Term
TERM_1
Z=15 + j * -20

INDQ
L1
L=1.002 nH

Term
TERM_2
Z=10 + j * -15

S-PARAMETERS

S_Param
SP1
Start=0 MHz
Stop=4000 MHz
Step=1 MHz

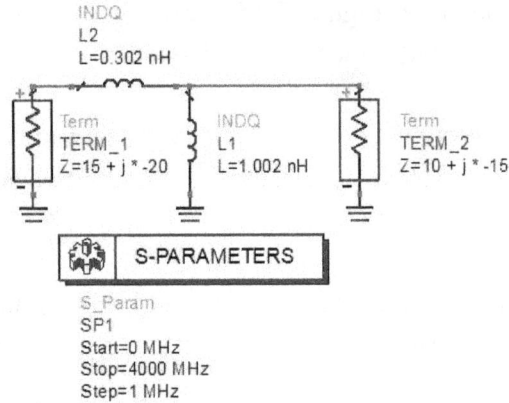

Figure 5-16 Simulation of the fourth matching L-network

Simulate the schematic and display S11 and S21 in a rectangular plot.

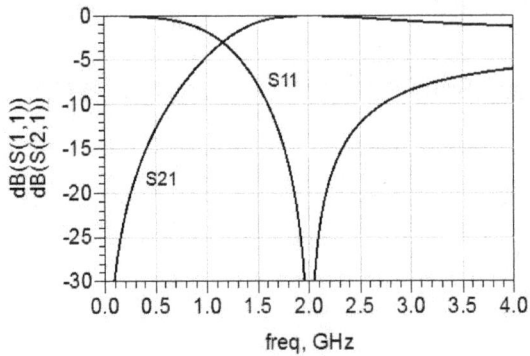

Figure 5-17 Response of the fourth matching L-network

The simulated response in Figure 5-17 shows that the matching L-network perfectly matches the load impedance to the source impedance at 2 GHz.

Matching a Complex Load to Real Source Impedance

The matching conditions for valid solutions are summarized in Table 5-1.

Case #	First Condition	Second Condition	# of Solutions	Equations Used
1	$R_L < R_S$	$R_L{}^2 + X_L{}^2 - R_L R_S > 0$	4	(5-21) to (5-24) (5-26) to (5-29)
2	$R_L < R_S$	$R_L{}^2 + X_L{}^2 - R_L R_S < 0$	2	(5-21) to (5-24)
3	$R_L > R_S$	N/A	2	(5-26) to (5-29)

Table 5-1 Impedance matching conditions and the number of solutions

The simplified version of these equations are given in the textbook.

Example 5.3-2A: Design a single L-network that will match a real source impedance $Z_0 = 50\ \Omega$ to a complex load impedance, $Z_L = 7 - j22\ \Omega$, at a frequency of 1 GHz.

Notice that for this example $R_L < Z_0$ and $R_L{}^2 + X_L{}^2 - R_L Z_0 = 183 > 0$, therefore, the matching network has four solutions.

First Solution: We use Equations (5-32) and (5-33) in the Appendix B to calculate the element values of the matching network.

1. Enter design parameters and normalized load impedance

Z0=50; RL=7; XL=-22; f=1000e6, r=RL/Z0, x=XL/Z0

2. Use Equations (5-32) and (5-33) in the textbook to calculate the matching element values

B1=sqrt((1-r)/r)/Z0X1=Z0*sqrt(r*(1-r))-x*Z0

LS1=X1/(2*pi*f)

CP2=B1/(2*pi*f)

The calculation results show that B1 = 0.05 and X1 = 39.349, therefore, the series element is an inductor L1= 6.263 nH and the shunt element is a capacitor C2=7.889 pF, as shown in Figure 5-18.

To plot the response of the matching network, create a new workspace in ADS and open a new schematic window. Insert the S_Params Template and add capacitor and inductor from the Lumped-Components palette. Connect the matching elements as shown in Figure 5-18.

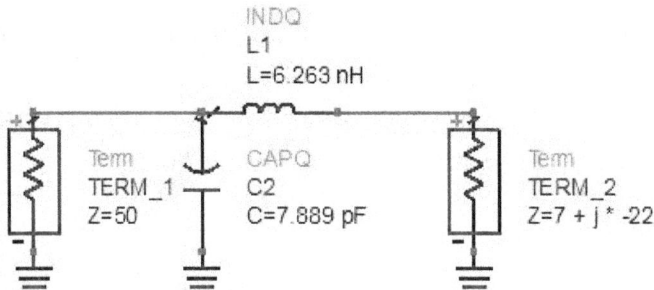

Figure 5-18 Schematic of the first matching L-network

Simulate the schematic and display S11 and S21 in a rectangular plot.

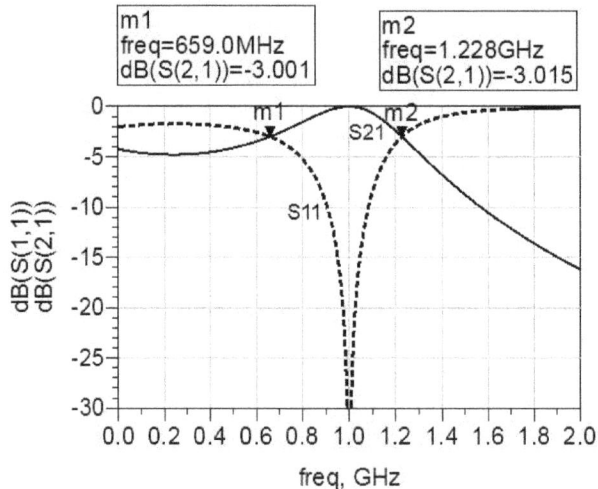

Figure 5-19 Response of the first matching L-network

The simulated response in Figure 5-19 shows that the matching L-network perfectly matches the load impedance to the source impedance at 1 GHz.

Example 5.3-2B: Design the second L-network that will match a real source impedance $Z_0 = 50\ \Omega$ to a complex load impedance, $Z_L = 7 - j22\ \Omega$, at a frequency of 1 GHz.

Second Solution: Use Equations (5-34) and (5-35) in the Appendix B to calculate the element values of the second matching L-network. The procedure follows.

1. Enter design parameters and normalized load impedance

Z0=50; RL=7; XL=-22; f=1000e6, r=RL/Z0, x=XL/Z0

2. Use Equations (5-34) and (5-35) in the textbook to calculate the matching element values

B2=-sqrt((1-r)/r)/Z0

X2=-Z0*sqrt(r*(1-r))-x*Z0

L1=X2/(2*pi*f)

L2=-1/(2*pi*f*B2)

The calculation results show that B2 = -o.05 and X2 = 4.651, therefore, the series element is an inductor L1= 0.74 nH and the shunt element is another inductor L2= 3.211 nH, as shown in Figure 5-20.

To plot the response of the matching network, create a new workspace in ADS and open a new schematic window. Insert the S_Params Template and add capacitor and inductor from the Lumped-Components palette. Connect the matching elements as shown in Figure 5-20.

INDQ
L1
L=0.74 nH

Term	INDQ	Term
TERM_1	L2	TERM_2
Z=50	L=3.211 nH	Z=7 + j * -22

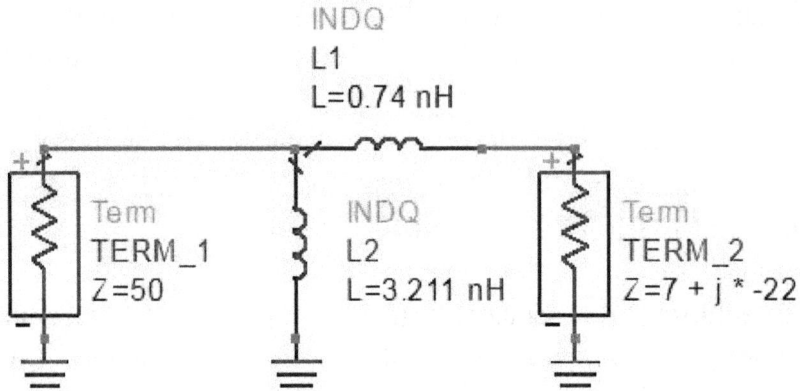

Figure 5-20 Schematic of the second matching L-network

Simulate the schematic and display S11 and S21 in a rectangular plot.

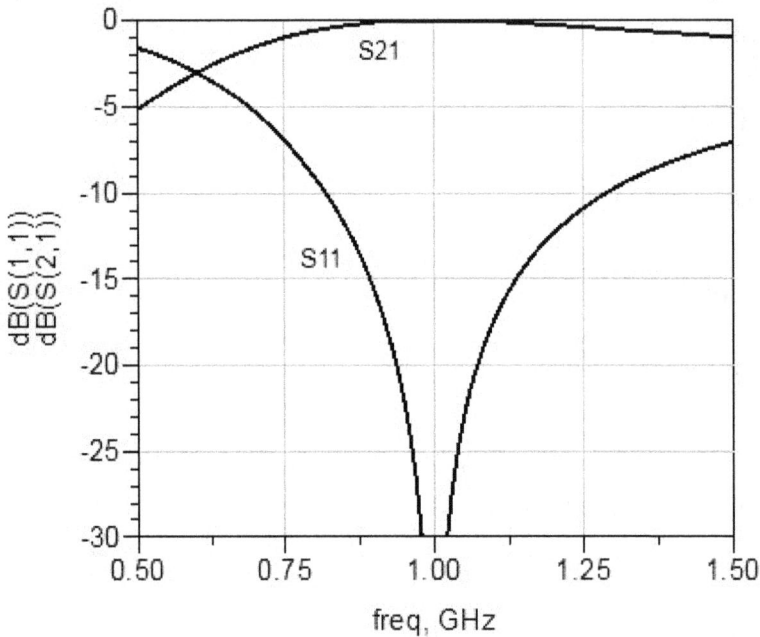

Figure 5-21 Response of the second matching L-network

The simulated response in Figure 5-21 shows that the matching L-network perfectly matches the load impedance to the source impedance at 1 GHz.

Example 5.3-2C: Design the third L-network that will match a real source impedance $Z_0 = 50\ \Omega$ to a complex load impedance, $Z_L = 7 - j22\ \Omega$, at a frequency of 1 GHz.

Third Solution: We use Equations (5-36) and (5-37) in the Appendix B to calculate the element values of the matching L-network. The procedure follows.

1. Enter design parameters and normalize the load impedance

Z0=50; RL=7; XL=-22; f=1000e6, r=RL/Z0, x=XL/Z0

2. Use Equations (5-36) and (5-37) in Matlab script to calculate the matching element values

B3=(x+sqrt(r*(r^2+x^2-r)))/(Z0*(r^2+x^2))

X3=Z0*sqrt((r^2+x^2-r)/r)

L2=X3/(2*pi*f)

L1=-1/(2*pi*f*B3)

The calculation results show that B3 = -0.032 and X3 = 36.154, therefore, the shunt element is an inductor L1= 5.008 nH and the series element is another inductor L2=5.754 nH, as shown in Figure 5-22.

To plot the response of the matching network, create a new workspace in ADS and open a new schematic window. Insert the S_Params Template and add capacitor and inductor from the Lumped-Components palette. Connect the matching elements as shown in Figure 5-22.

INDQ
L2
L=5.754 nH

Term
TERM_1
Z=50

INDQ
L1
L=5.008 nH

Term
TERM_2
Z=7 + j * -22

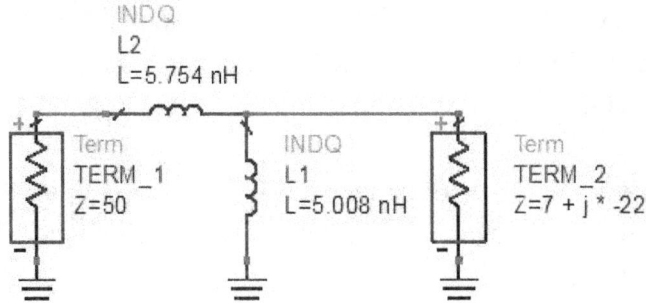

Figure 5-22 Schematic of the third matching L-network

Simulate the schematic from 500 to 1500 MHz and display the input return loss, S11, and insertion loss, S21, in dB.

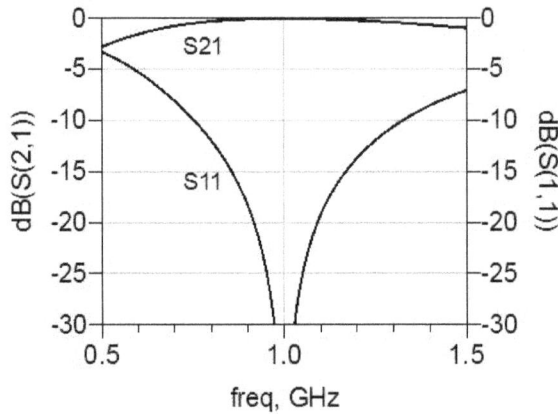

Figure 5-23 Response of the third matching L-network

The simulated response in Figure 5-23 shows that the matching L-network perfectly matches the load impedance to the source impedance at 1 GHz.

Example 5.3-2D: Design the fourth L-network that will match a real source impedance $Z_0 = 50 \ \Omega$ to a complex load impedance, $Z_L = 7 - j22 \ \Omega$, at a frequency of 1 GHz.

Fourth Solution: For the fourth solution use Equations (5-38) and (5-39) in the Appendix B to calculate the element values of the matching network. The procedure follows.

1. Enter design parameters and normalize the load impedance

Z0=50; RL=7; XL=-22; f=1000e6, r=RL/Z0, x=XL/Z0

2. Use Equations (5-38) and (5-39) to calculate the matching element values

B4=(x-sqrt(r*(r^2+x^2-r)))/(Z0*(r^2+x^2))

X4=-Z0*sqrt((r^2+x^2-r)/r)
C2=-1/(2*pi*f*X4)

L1=-1/(2*pi*f*B4)

The solution in the second configuration shows that B4 = -0.051 and X1 = -36.154, therefore, the shunt element is an inductor L1= 3.135 nH and the series element is a capacitor C2=4.402 pF.

To plot the response of the matching network, create a new workspace in ADS and open a new schematic window. Insert the S_Params Template and add capacitor and inductor from the Lumped-Components palette. Connect the matching elements as shown in Figure 5-24.

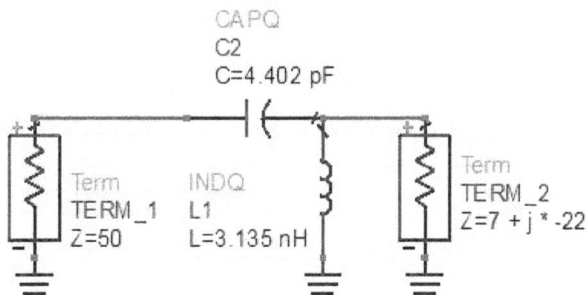

Figure 5-24 Schematic of the fourth matching L-network

Simulate the schematic and display S11 and S21 in a rectangular plot.

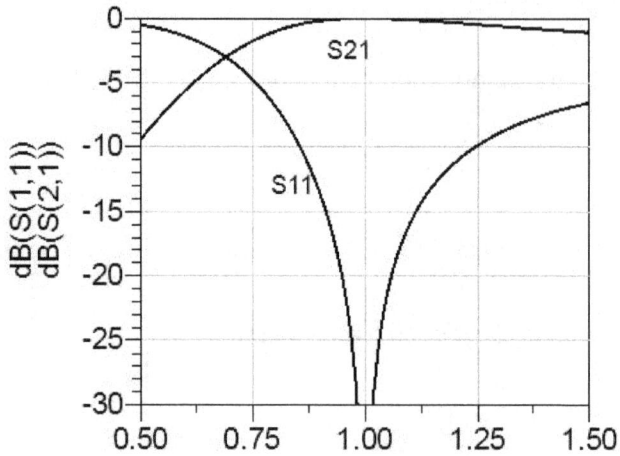

Figure 5-25 Response of the fourth matching L-network

The simulated response in Figure 5-25 shows that the matching L-network perfectly matches the load impedance to the source impedance at 1 GHz.

Matching a Real Load to a Real Source Impedance

When source and load impedances are both real the first matching configuration of Figure 5-7 is redrawn in Figure 5-26.

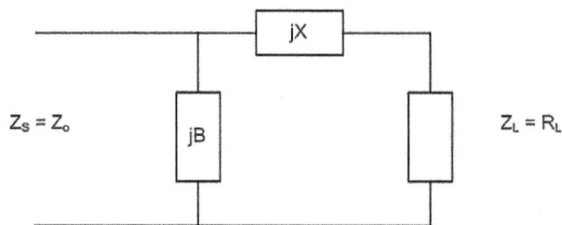

Figure 5-26 First matching configuration with $X_L = X_S = 0$

The two conditions in this case are summarized in Table 5-2.

Case No	Condition	Solutions	Equations
1	$r < 1$	2	(5-42) to (5-45)
2	$r > 1$	2	(5-47) to (5-50)

Table 5-2 Impedance matching conditions and the number of solutions

Example 5.3-3A: Design an L-network to match a 10 Ω load to a 50 Ω source resistor at 500 MHz.

Because the load resistor is smaller than the source resistor, the example has two solutions given in Equations (5-42) through (5-45).

First Solution: For the first solution use Equations (5-42) and (5-43) in the Appendix B to calculate the element values of the matching L-network. The procedure follows.

1. Enter design parameters and normalized load impedance

Z0=50; RL=10; f=500e6, r=RL/Z0,

2. Use Equations (5-42) and (5-43) in Matlab script to calculate the matching element values

B1=sqrt((1-r)/r)/Z0

X1=Z0*sqrt(r*(1-r))

L1=X1/(2*pi*f)

C2=B1/(2*pi*f)

The calculation results show that B1 = 0.04 and X1 = 20, therefore, L1=6.366 nH and C2=12.73 pF.

To generate the schematic and plot the response of the matching network, create a new workspace in ADS and open a new schematic window. Insert the S_Params Template and add capacitor and inductor from the Lumped-Components palette. Connect the matching elements as shown in Figure 5-27.

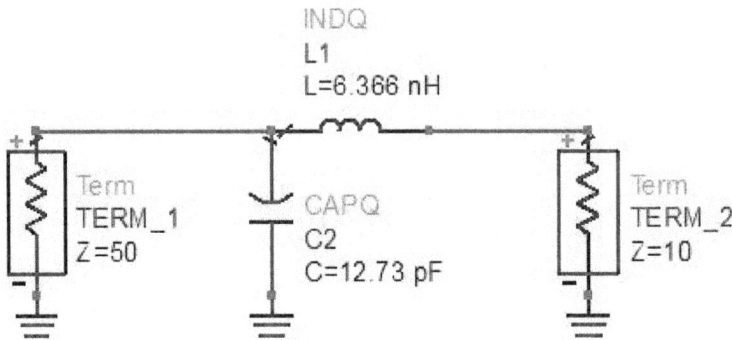

Figure 5-27 Schematic of the first matching L-network

Simulate the schematic and display S11 and S21 in a rectangular plot.

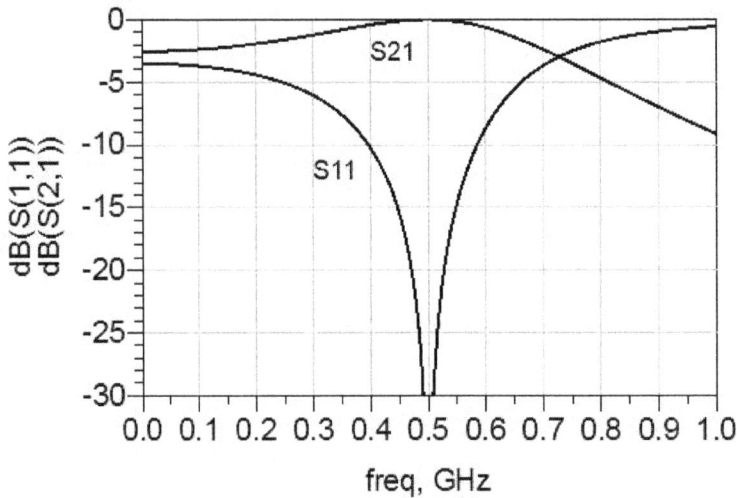

Figure 5-28 Response of the first matching L-network

The simulated response in Figure 5-33 shows that the matching L-network perfectly matches the load impedance to the source impedance at 0.5 GHz.

Example 5.3-3B: Design the second L-network to match a 10 Ω load to a 50 Ω source resistor at 500 MHz.

Second Solution: To design the second matching L-network, use Equations (5-44) and (5-45) in the Appendix B to calculate the matching element values.

1. Enter design parameters and normalize the load impedance

Z0=50; RL=10; f=500e6, r=RL/Z0,

2. Calculate matching element values

B2=-sqrt((1-r)/r)/Z0

X2=-Z0*sqrt(r*(1-r))

C1=-1/(2*pi*f*X2)

L2=-1/(2*pi*f*B2)

The calculation results show that B2 = -0.04 and X2 = -20, therefore, the series element is a capacitor, C=15.92 pF, and the shun element is an inductor, L= 7.958 nH. To generate the schematic and plot the response of the matching network, create a new workspace in ADS and open a new schematic window. Insert the S_Params Template and add capacitor and inductor from the Lumped-Components palette. Connect the matching elements as shown in Figure 5-29.

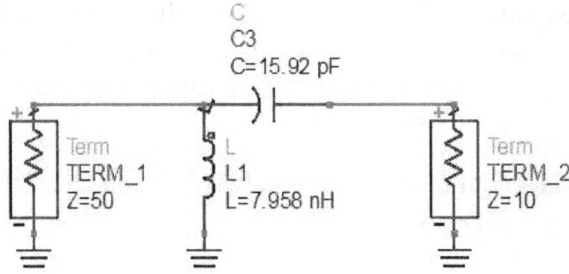

Figure 5-29 Schematic of the second matching L-network

Simulate the schematic and display S11 and S21 in a rectangular plot.

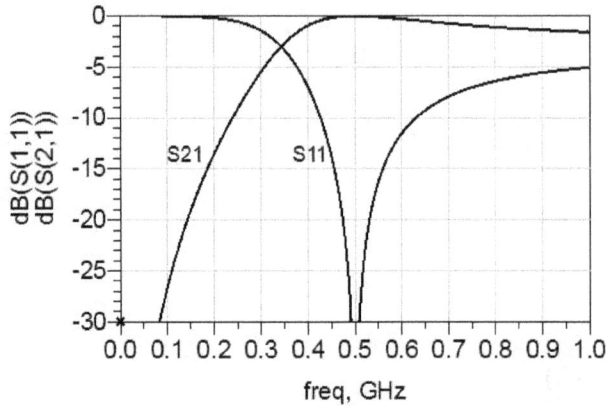

Figure 5-30 Response of the second matching L-network

The simulated response in Figure 5-30 shows that the matching L-network perfectly matches the load impedance to the source impedance at 0.5 GHz.

Introduction to Broadband Matching Networks

In the previous sections, the L-section matching networks achieved an impedance match at a fixed frequency capable of producing a 20 dB return loss over a narrow fractional bandwidth of less than 20%. Broadband networks are generally considered to have greater than 20% fractional bandwidths. In this section it is demonstrated that the bandwidth of a matching network can be increased by cascading L-networks. It is demonstrated that by cascading L-networks of equal Q factor, the bandwidth of a network can be increased. The design of equal-Q matching networks is based on the selection of intermediate, or virtual, resistors not necessarily 50 Ω, and then matching the load and source impedance to the virtual resistors. Successively adding additional L-networks of equal Q will continue to extend the bandwidth of the overall circuit.

Analytical Design of Broadband Matching Networks

This section demonstrates the importance of the selection of the proper intermediate network resistance that will result in the best broadband return loss. In example 5.4-1 the complex source and load impedance are matched to one specific intermediate resistor thus creating two, equal-Q, L-networks. The purpose is to show that this method provides a broader matching bandwidth at 20 dB return loss compared to the case when we chose a different resistor value. The Q of each L-network is the loaded Q factor defined by the source and load resistance ratio.

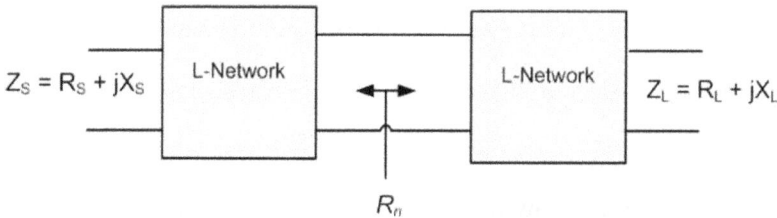

Figure 5-31 Cascaded L-networks with intermediate resistance, R_n

Example 5.4-1: A transmitter operates over a frequency range of 835 MHz to 1200 MHz. At its center frequency of 1 GHz the source impedance is $Z_S = 55 + j10\ \Omega$ while the antenna input impedance is $Z_L = 20 + j15\ \Omega$. Analytically design a cascade of two L-networks that matches the transmitter output impedance to the antenna impedance across the full transmitter bandwidth.

Solution: The first step in the solution is to calculate the intermediate resistor, R_n. The optimum intermediate resistor value is equal to square root of the product of the real part of the source and load impedance.

$$R_n = \sqrt{(55)(20)} = 33.166 \ \Omega$$

We use MATLAB scripting and the Equations (5-32) and (5-33) in the Appendix B to calculate the element values of the load matching network.

1. Enter design parameters and normalized load impedance

Z0=33.166; RL=20; XL=15; f=1000e6, r=RL/Z0, x=XL/Z0

2. Use Equations (5-32) and (5-33) in the Appendix B to calculate the matching element values

B1=sqrt((1-r)/r)/Z0

X1=Z0*sqrt(r*(1-r))-x*Z0

LS1=X1/(2*pi*f)

CP2=B1/(2*pi*f)

The calculation results show that the series element is an inductor, L1 = 0.195 nH, and the shunt element is a capacitor, C1 = 3.893 pF.

To plot the response of the matching network, create a new workspace in ADS and open a new schematic window. Insert the S_Params Template and add capacitor and inductor from the Lumped-Components palette. Connect the matching elements as shown in Figure 5-32.

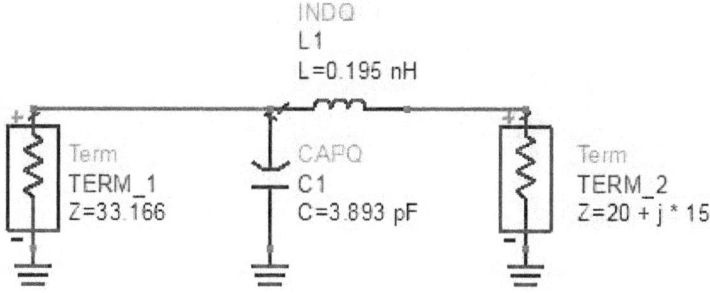

Figure 5-32 Schematic of the matching L-network

Simulate the schematic and display S11 and S21 in a rectangular plot.

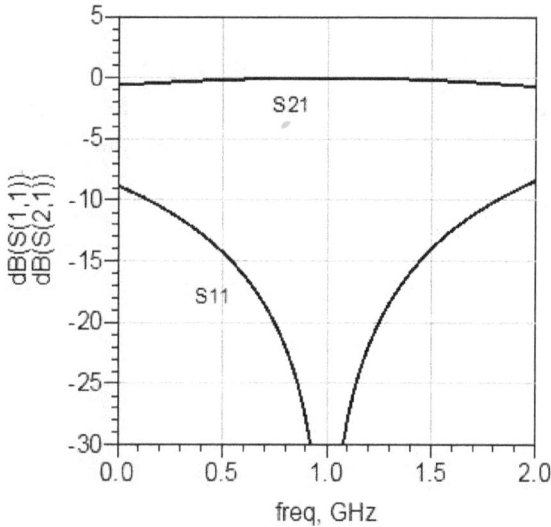

Figure 5-33 Response of the matching L-network

To match the source impedance to the intermediate resistor, we use Equations (5-36) and (5-37) in the Appendix B and calculate the element values of the second matching network, as follows.

1. Enter design parameters and normalized load impedance

Z0=33.166; RL=55; XL=10; f=1000e6, r=RL/Z0, x=XL/Z0

2. Use Equations (5-36) and (5-37) in the Appendix B to calculate the source matching element values

B3=(x+sqrt(r*(r^2+x^2-r)))/(Z0*(r^2+x^2))

X3=Z0*sqrt((r^2+x^2-r)/r)

LS1=X3/(2*pi*f)

CP2=B3/(2*pi*f)

The calculation results show that ten shunt element is a capacitor, C1= 2.875 pF, and the series element is an inductor, L1=4.458 nH.

Create a new workspace in ADS and open a new schematic window. Insert the S_Params Template and add capacitor and inductor from the Lumped-Components palette. Connect the matching elements as shown in Figure 5-34.

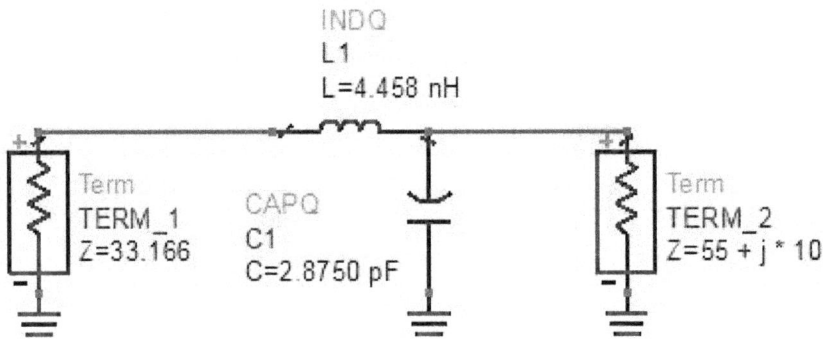

Figure 5-34 Schematic of the matching L-network

To plot the response of the matching network, Simulate the schematic

and display S11 and S21 in a rectangular plot.

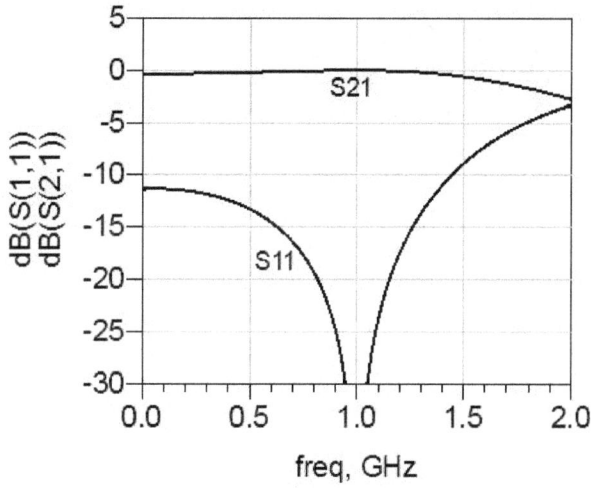

Figure 5-35 Response of the matching L-network

Finally, we cascade the two matching L-networks as shown in Figure 5-36.

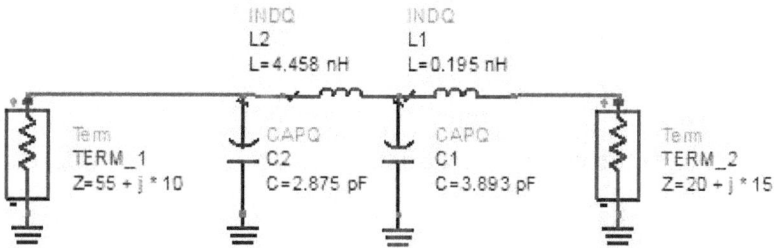

Figure 5-36 Schematic of the Cascaded L-networks

Simulate the schematic and display S11 and S21 in a rectangular plot.

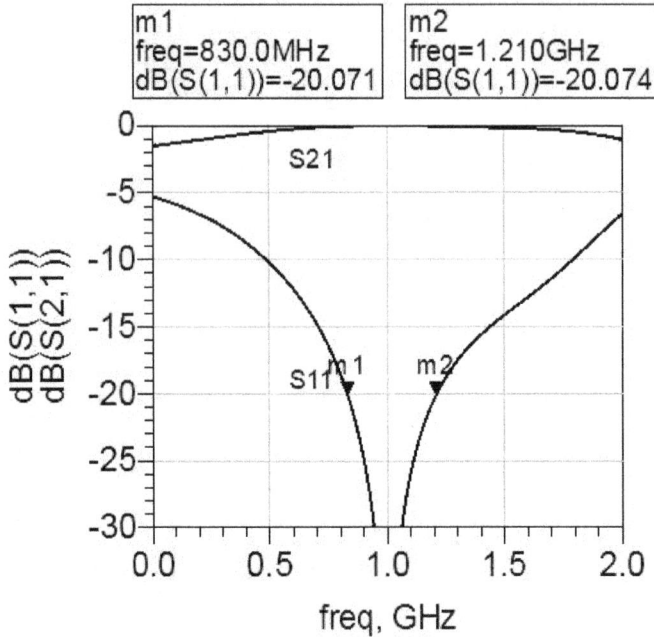

Figure 5-37 Response of the Cascaded L-networks

The simulated response in Figure 5-42 shows that the cascaded network perfectly matches the complex load impedance to the source impedance at 1 GHz.

Notice that the matching bandwidth at 20 dB return loss is:

$$BW_{20\,dB\ return\ loss} = 1210 - 830 = 380\ MHz$$

Therefore, the corresponding fractional bandwidths at 20dB return loss is:

$$FBW_{20dB\ return\ loss} = \frac{380}{\sqrt{(1210)(830)}} = 0.379 = 37.9\ \%$$

The Q factor of the each L-network is calculated by the following equation.

$$Q = \frac{1}{2}\sqrt{\frac{R_2}{R_1} - 1} \qquad\qquad R_2 > R_1$$

where R_2 and R_1 are the input and output resistors of the L-network.

For Example 5.4-1 the Q of the source and load matching networks are:

$$Q_{source} = Q_{load} = \frac{1}{2}\sqrt{\frac{33.166}{20} - 1} = \frac{1}{2}\sqrt{\frac{55}{33.166} - 1} = 0.405$$

Example 5.4-2: Redesign the matching network by matching the source and load impedances to an intermediate resistance value of 50 .Ω Compare the fractional bandwidth and Q factors with values for Example 5.4-1.

Solution: To redesign the matching network, change the intermediate resistor to 50 Ω and recalculate the matching element values. The schematics are shown in Figure 5-38.

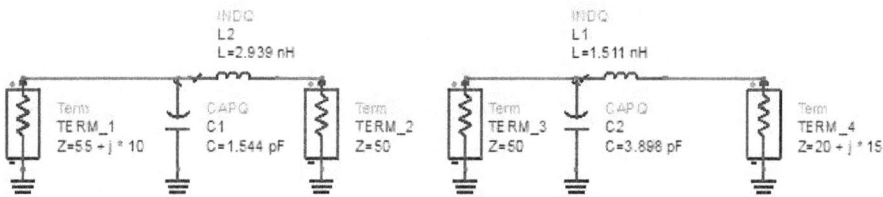

Figure 5-38 Load and source impedance matched to 50 Ω

Notice that the matching element values in both schematics have changed due to the change in the intermediate resistor. The cascaded schematic and simulated response are shown in Figures 5-39 and 5-40.

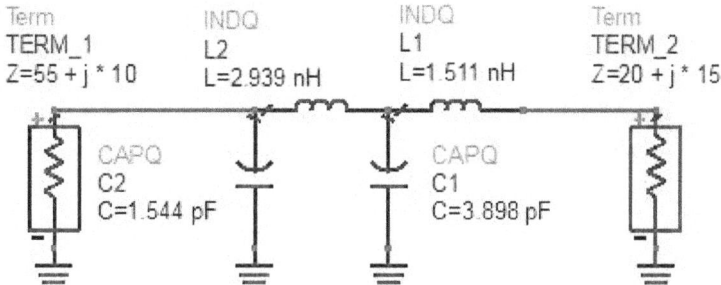

Figure 5-39 schematic of the cascaded matching L-networks

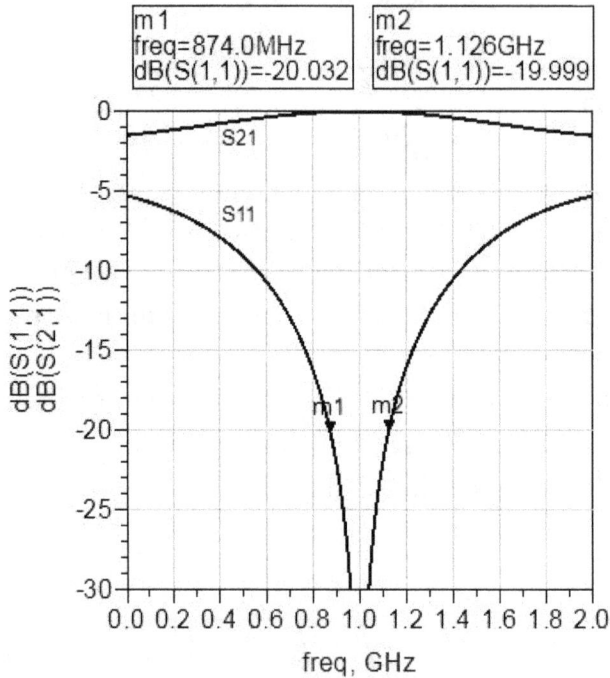

Figure 5-40 Response of the cascaded matching L-networks

The simulated response in Figure 5-40 shows that the matching L-network perfectly matches the load impedance to the source impedance at 0.5 GHz.

The marker frequencies at 20 dB return loss shows that the bandwidth is:

$$BW_{20\,dB\ return\ loss} = 1126 - 874 = 252\ MHz$$

Therefore, the corresponding fractional bandwidths at 20dB return loss is:

$$FBW_{20\,dB\ return\ loss} = \frac{252}{\sqrt{(1126)(874)}} = 0.253 = 25.4\ \%$$

A comparison of the fractional bandwidths for the two examples show that the matching network in Figure 5-37 provides over 50 % more matching bandwidth, at 20 dB return loss, than the matching network in Figure 5-40.

We can show that the lesser bandwidth is due to the fact that the source and load matching networks in the second example do not have the same Q factors, as shown by the following calculations:

$$Q_{source} = \frac{1}{2}\sqrt{\frac{55}{50}-1} = 0.158$$

$$Q_{load} = \frac{1}{2}\sqrt{\frac{50}{20}-1} = 0.612$$

The equal Q factors in Example 5.4-1 cause the cascaded matching network to transfer more power to the load while the unequal Q factors in example 5.4-2 cause the matching network to have less power transferred.

Broadband Impedance Matching Using N-Cascaded L-Networks

In Example 5.4-1 we showed that cascading two equal-Q L-networks convert a narrowband matching L-network into a broadband matching network. Because the Q factor is related to the ratio of the source and load resistance on each L-network, the Q can be further reduced by applying multiple L-networks in cascade. In practice as the number of L-networks becomes greater than five, the element values become difficult to physically realize. In this section we show that by using a network of four equal-Q L-networks in cascade, we can lower the individual Q factor and provide an even greater matching bandwidth for the Example 5.4-1.

Example 5.4-3: Redesign the matching network of Example 5.4-1 with four equal-Q L-networks. Compare the Q and bandwidth of this example with Example 5.4-1.

Solution: This example uses the same steps developed for Example 5.4-1. The general equation for the calculation of any number of intermediate resistors is given by the following Equation.

$$R_n = R_S(r)^{n/N} \qquad\qquad n = 1, 2, 3, to\ N-1$$

where:

      R_n is the intermediate resistor values
      R_S is the source resistor value
      r is the normalized load resistor, $r = R_L/R_S$
      N is the number of cascading networks

Using the above equation, calculate the intermediate resistors for N = 4.

1. Enter design parameters and normalize the load impedance

RS=55; RL=20; r=RL/RS; N=4

2. Use Matlab script to calculate intermediate resistor values

R1=RS*r^(1/N

R2=RS*r^(2/N

R3=RS*r^(3/N

The calculation results show that R1=42.71 Ω, R2=33.166 Ω, and R3=25.755 Ω.

Now design 4 matching L-network from the load toward the source.

1. Enter design parameters and normalized load impedance

R3=25.755; RL=20; XL=15; f=1000e6, r=RL/R3, x=XL/R3

2. Use Equations (5-32) and (5-33) to design the first L-network

B1=sqrt((1-r)/r)/R3

X1=R3*sqrt(r*(1-r))-x*R3

CS1=-1/(2*pi*f*X1)

CP2=B1/(2*pi*f)

The calculation results show that C1= 37.26 pF and C2= 3.315 pF, as shown in Figure 5-41.

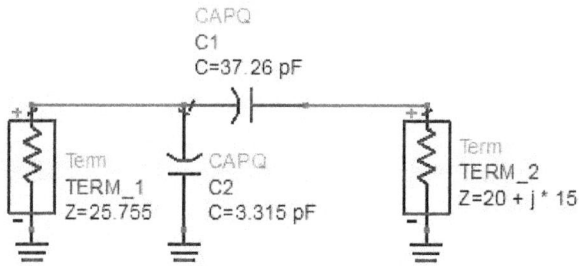

Figure 5-41 Schematic of the first matching L-network

Similarly design the second matching L-network between R3 and R2.

1. Enter design parameters and normalized load impedance

R2=33.166; RL=25.755; f=1000e6, r=RL/R2

2. Design the second matching L-network

B1=sqrt((1-r)/r)/R2

X1=R2*sqrt(r*(1-r))

LS1=X1/(2*pi*f)

CP2=B1/(2*pi*f)

The calculation results show that L= 2.199 nH and C1= 2.574 pF, as shown in Figure 5-42.

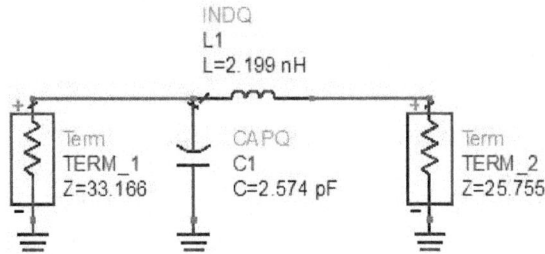

Figure 5-42 Schematic of the second matching L-network

Next, design the third matching L-network, between R2 and R1.

1. Enter design parameters and normalized load impedance

R1=42.71; RL=33.166; f=1000e6, r=RL/R1

2. Design the third matching L-network

B1=sqrt((1-r)/r)/R1

X1=R1*sqrt(r*(1-r))

LS1=X1/(2*pi*f)

CP2=B1/(2*pi*f)

The calculation results show that L1= 832 nH and C1=1.999 pF as shown in Figure 5-43.

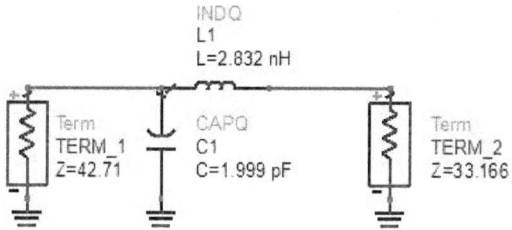

Figure 5-43 Schematic of the third matching L-network

Finally, design the fourth matching L-network between R1 and source

1. Enter design parameters and normalized load impedance

R1=42.71; RL=55; XL=10; f=1000e6, r=RL/Z0, x=XL/Z0

2. Design the forth matching L-network

B3=(x+sqrt(r*(r^2+x^2-r(R1*(r^2+x^2

X3=R1*sqrt((r^2+x^2-r/r

LS1=X3/(2*pi*f

CP2=B3/(2*pi*f

The calculation results show that L1=3.907 nH and C1= 2.119 pF.

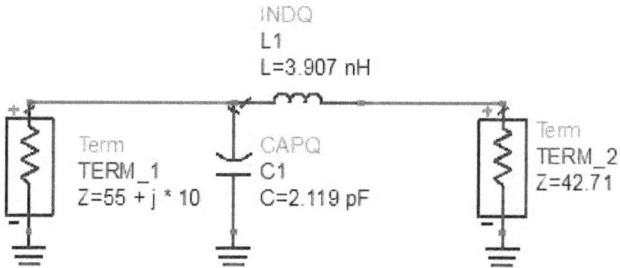

Figure 5-44 Schematic of the fourth matching L-network

Cascade the four L-networks as shown in Figure 5-45.

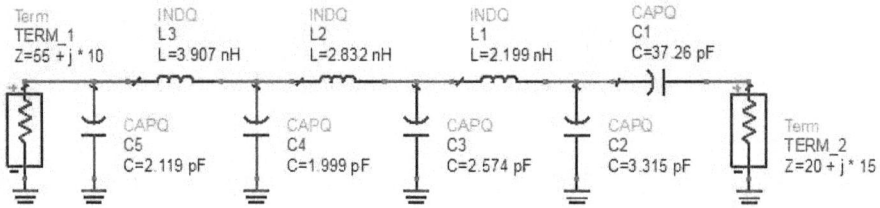

Figure 5-45 Schematic of the four cascaded matching L-networks

The simulated response shows that the matching bandwidth at 20 dB return loss is:

$$BW_{20dB \; return \; loss} = 1334 - 821 = 513 \; MHz$$

Therefore, the fractional bandwidth at 20 dB return loss is:

$$FBW_{20dB \; return \; loss} = \frac{513}{\sqrt{(1334)(821)}} = 0.490 = 49.0\%$$

The simulated response is shown in Figures 5-46.

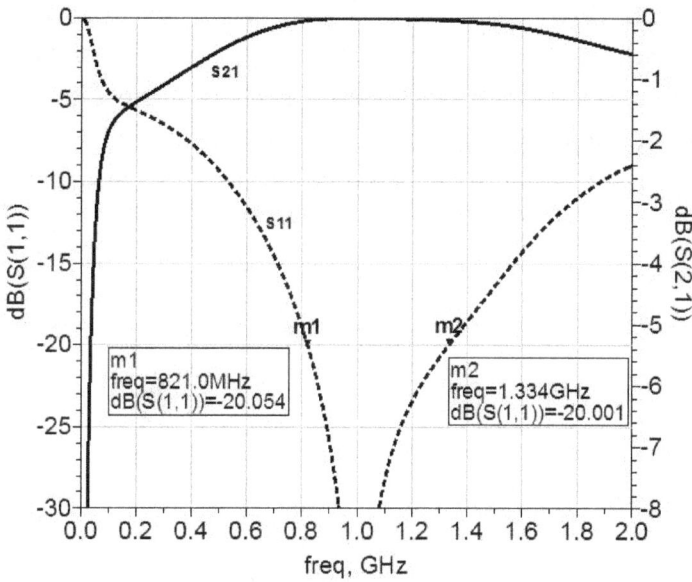

Figure 5-46 Response of the 4 cascaded matching L-networks

The simulated response in Figure 5-46 shows that the matching L-network perfectly matches the load to the source impedance at 1 GHz.

The comparison between the fractional bandwidth of this example with the fractional bandwidth of example 5.4-1 shows that cascading four equal-Q matching L-networks increases the fractional bandwidth by more than 93 % over the network with only two equal-Q matching L-networks.

As we stated earlier, the reason for wider matching bandwidth is that all matching L-networks have the same Q factor as calculated here.

$$Q_{4L} = \frac{1}{2}\sqrt{\frac{55}{42.71}-1} = \frac{1}{2}\sqrt{\frac{42.71}{33.166}-1} = \frac{1}{2}\sqrt{\frac{33.166}{25.755}-1} = \frac{1}{2}\sqrt{\frac{25.755}{20}-1} = 0.268$$

Limitations of Broadband Matching

From the previous example we can deduce that the Q of the complex source and load impedances have an impact on the maximum bandwidth over which a good impedance match, low return loss, can be achieved. There is a finite limit on the achievable return loss for a given load impedance and circuit bandwidth known as Fano's limit [1]. Fano's Limit is the optimum reflection coefficient that can be achieved with a given load impedance. It is a theoretical limit that considers an infinite number of lossless matching elements. Fano's limit is defined by the Equation.

$$|\Gamma| = e^{\frac{-\pi Q_L}{Q_{UL}}}$$

Where:

- Q_{UL} = Ratio of the reactance to resistance of the load

- Q_L = the loaded Q of the network or the ratio of the center frequency of the network divided by the 3 dB bandwidth.

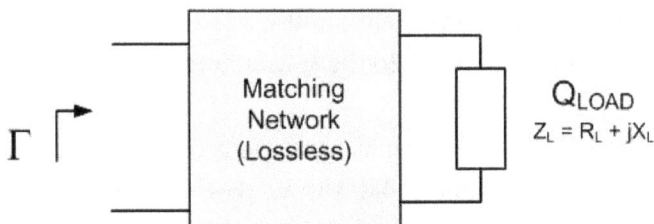

Figure 5-47 General configuration of impedance matching network

Example of Fano's Limit Calculation

In the following example we apply the Fano's limit to calculate the optimum reflection coefficient for a given network.

Example 5.7-1 Use the ADS Impedance Matching Utility to match a source impedance of 100 Ω to a load that is comprised of a 500 Ω resistor in parallel with a 2 pF capacitance as shown in Figure 5-48. It is required that a 20 dB return loss and greater than -1 dB insertion loss be achieved over a frequency range of 60 MHz to 160 MHz.

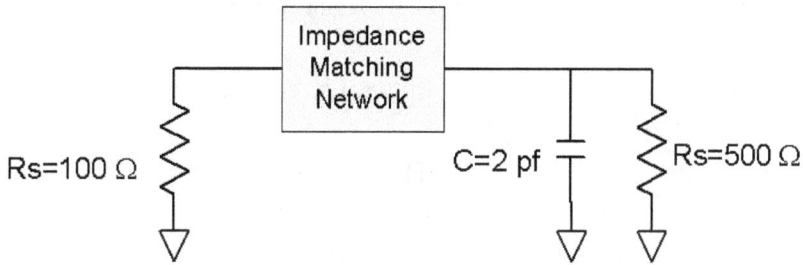

Figure 5-48 Matching the load to the 100 Ω source resistor

Solution: The Impedance Matching Utility in ADS provides SmartComponents and Matching Assistant for the design and simulation of impedance matching networks. The procedure follows.

1. Create a new workspace in ADS and open a new schematic window.

2. From the schematic window, click Tools > Impedance Matching to open the Impedance Matching Utility Control window.

3. From the Utility window toolbar, click View > Component Palette to place the matching SmartComponents on the schematic window

4. Click DA_LCBandpassMatch to select the component.

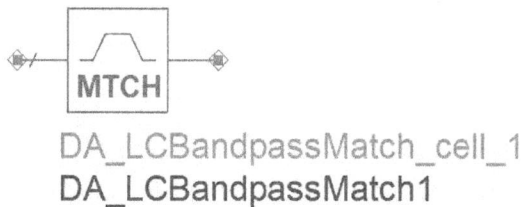

Figure 5-49 Bandpass filter SmartComponent

5. In the Control window. Select the DA_LCBandpassMatch component from the SmartComponent drop-down list.

6. Select the **Matching Assistant** tab and start editing the filter parameters

7. From the Source Impedance drop-down list box, select Resistive and set R=100 Ohm

8. From the Load Impedance drop-down list box, select Parallel RC and set R=500 Ohm and C=2 pF

9. For a center frequency of 100 MHz select the lower passband corner frequency Fp1= 61 MHz and the upper passband corner frequency Fp2=164 MHz, as shown in Figure 5-50

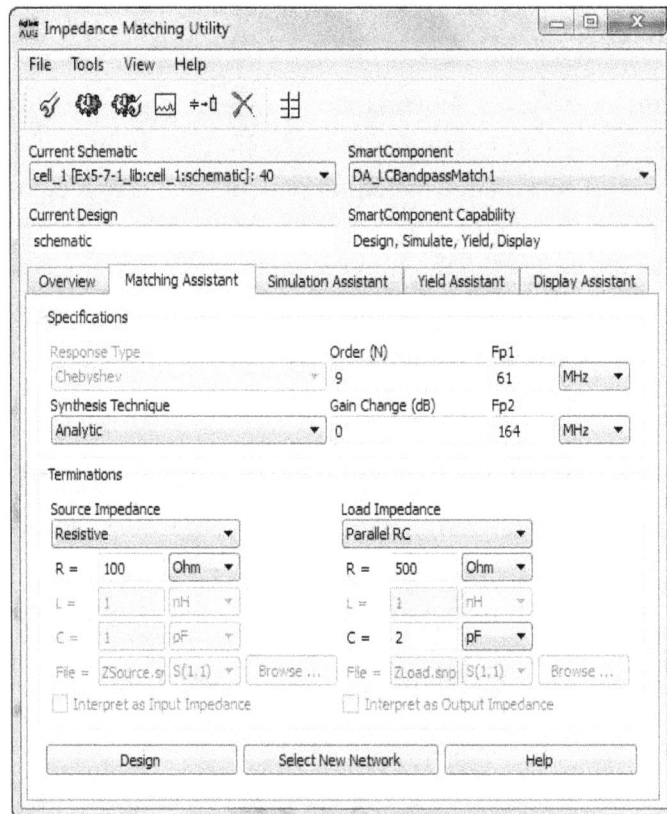

Figure 5-50 Impedance Matching Utility Control window

10. Increase the order of the bandpass matching network until the design requirements are met. Figure 5-51 shows the matching network.

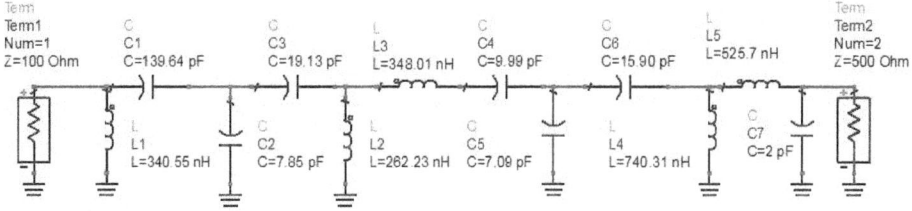

Figure 5-51 Schematic of the impedance matching network

The Impedance Matching SmartComponent with design parameters is shown in Figures 5-52.

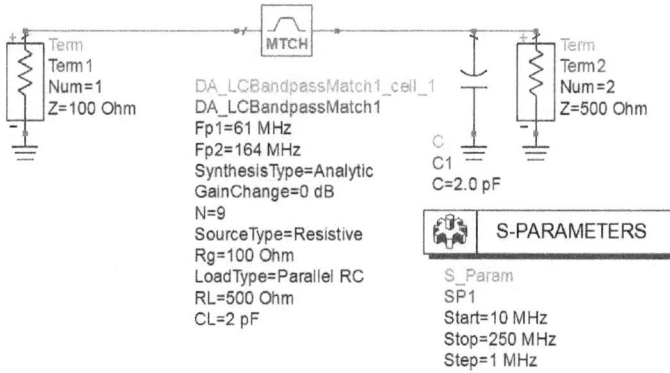

Figure 5-52 Bandpass impedance matching design

Simulate the schematic in Figure 5-52 to get the following response.

Forward Transmission, dB

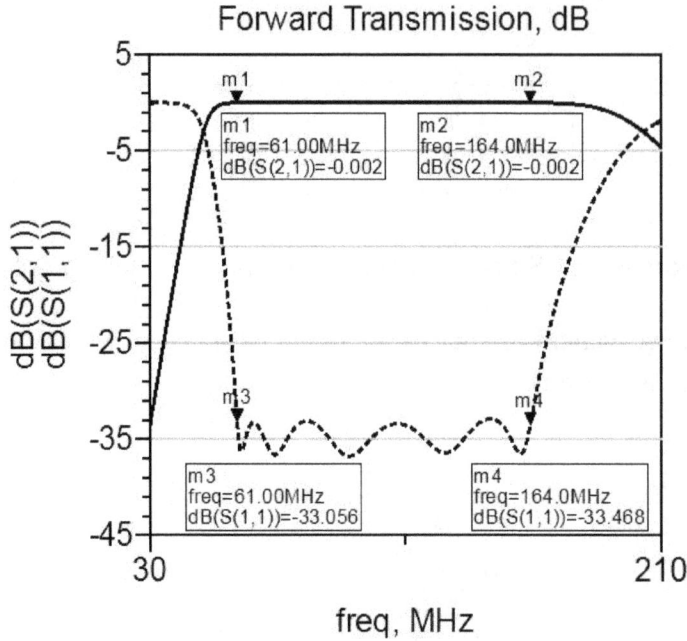

Figure 5-53 Response of the matching network (N = 9)

The marker readings m3 and m4 in Figure 5-53 show that the design requirements at 61 and 164 MHz are fully met.

Effect of Finite Q on the Matching Networks

To complete the matching circuit design, the components need to be converted to their physical equivalents by accounting for the finite Q factor of the capacitors and inductors. This can be accomplished by using either S parameter or Modelithics models for the elements. A quick approach is to model each element with a Q factor that closely approximates the physical components. In this design we'll demonstrate a technique to assign a common variable name representing the Q factor of each component. This allows the evaluation of the necessary Q factor of the components required to meet the specifications of the matching network. The variable name, Qcap, is assigned to all capacitors in the matching network. Similarly the variable name, Qind, is assigned to all of the inductors. The frequency at which the Q factor is defined is also defined as a variable, Qf. This allows the Q factor of all inductors and all capacitors to be tuned simultaneously.

Example 5.7-2 Assign Qcap and Qind values to capacitors and inductors in Figure 5-54. It is required that a 20 dB return loss and greater than -1 dB insertion loss be achieved over a frequency range of 60 MHz to 160 MHz

Solution: The network response with the initial Qcap = 150 and Qind = 100 values is shown in Figure 5-55. We can see that the insertion loss (>-1dB) and return loss (<20 dB) specifications have been met.

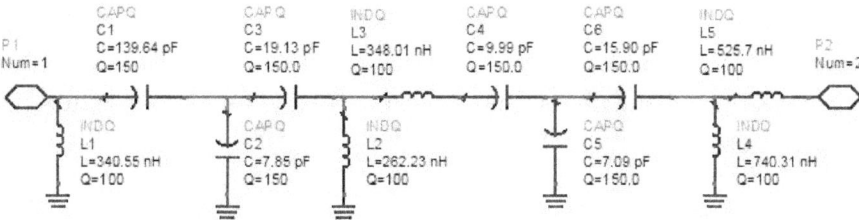

Figure 5-54 Schematic of the cascaded matching L-networks
(Qcap=150 and Qind=100)

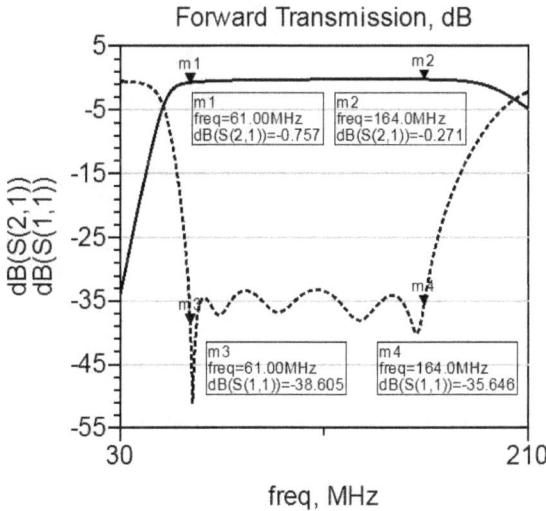

Figure 5-55 Response with Qcap=150 and Qind=100

As Figure 5-56 shows, a capacitor Q of 50 and an inductor Q of 40 results in an insertion loss that falls short of meeting the insertion loss specification.

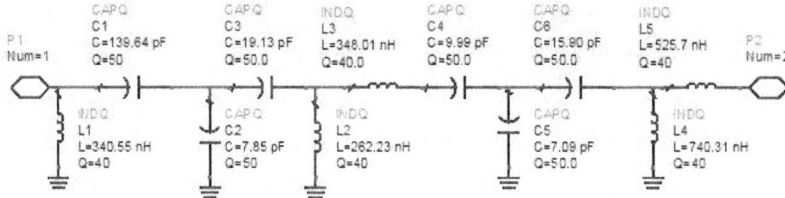

Figure 5-56 Schematic of the cascaded matching L-networks (Qcap=50 and Qind=40)

The network response with the initial Qcap = 50 and Qind = 40 values is shown in Figure 5-57.

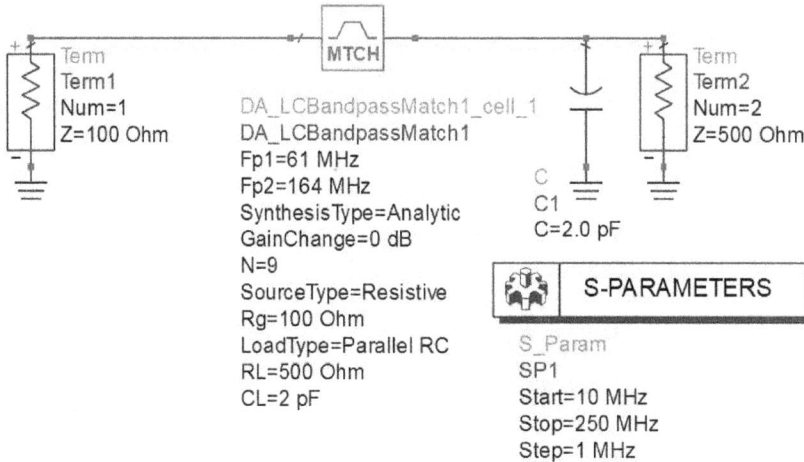

Figure 5-57 Bandpass impedance matching design

The response with Qcap=50 and Qind=40 is shown in Figure 5-65.

Forward Transmission, dB

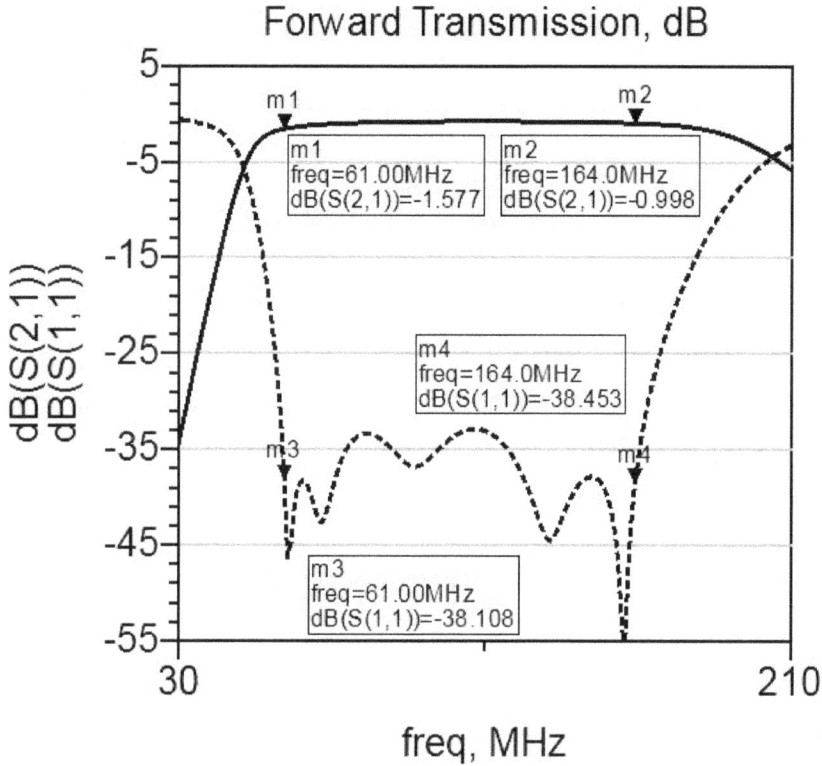

m1
freq=61.00MHz
dB(S(2,1))=-1.577

m2
freq=164.0MHz
dB(S(2,1))=-0.998

m4
freq=164.0MHz
dB(S(1,1))=-38.453

m3
freq=61.00MHz
dB(S(1,1))=-38.108

freq, MHz

Figure 5-55 Response with Qcap=50 and Qind=40

References and Further Reading

[1]   Ali A. Behagi, *RF and Microwave Circuit Design,* A Design Approach Using (**ADS**). Techno Search, Ladera Ranch, CA 92694. August 2015.

[2]   Keysight Technologies, Manuals for Advanced Design System, *ADS 2016.01 Documentation Set*, EEsof EDA Division, Santa Rosa, California www.keysight.com

[3]   Guillermo Gonzales, *Microwave Transistor Amplifiers – Analysis and Design*, Second Edition, Prentice Hall Inc., Upper Saddle River, NJ.

[4]   Randy Rhea, *The Yin-Yang of Matching: Part 2 – Practical Matching Techniques*, High Frequency Electronics, March 2006

[5]   Steve C. Cripps, *RF Power Amplifiers for Wireless Communications*, Artech House Publishers, Norwood, MA. 1999.

[6]   David M. Pozar, *Microwave Engineering*, Third Edition, John Wiley & Sons, New York, 2005

[7]   R. Ludwig, P. Bretchko, *RF Circuit Design*, Theory and Applications, Prentice Hall, Upper Saddle River, NJ, 2000

[8]   Jerry Sevick, *Transmission Line Transformers*, The American Radio Relay League, Newington, CT., 1990.

Problems

5-1.   For a load of 75 Ω with a 10 pF series capacitance we want to have maximum power transfer at 1 GHz. Find a source inductance that cancels the reactance of the load at this frequency.

5-2.   Analytically design all the L-networks that will match a complex load impedance, $Z_L = 15 - j10 \ \Omega$, to a complex source impedance, $Z_S = 25 + j20 \ \Omega$, at a frequency of 1 GHz. Verify all the solutions by plotting the response of the matching networks.

5-3.   Analytically design all the L-networks that will match a source impedance, $Z_S = 30 \ \Omega$, to a complex load impedance, $Z_L = 25 + j20 \ \Omega$, at a frequency of 1 GHz. Verify all the solutions by plotting the response of the matching networks.

5-4.   Analytically design all the L-networks that will match a load impedance, $Z_L = 15 \ \Omega$, to a source impedance, $Z_S = 75 \ \Omega$, at a frequency of 1.5 GHz. Verify all the solutions by plotting the response of the matching networks.

5-5. The output impedance of a transmitter operating at a frequency of 2 GHz is: $Z_S = 30 + j10$ Ω. Analytically design all the matching L-networks such that the maximum power is delivered to the antenna whose input impedance is $Z_L = 10 + j15$ Ω. Compare the matching bandwidth at 20 dB return loss with the method of using 50 Ω interconnecting line.

5-6. Redesign the network of Problem 5-5 by matching the source and load impedances to a 50 Ω line. Compare fractional bandwidth and Q factors for examples 5.5 and 5.6.

5-7. Redesign the matching network in Example 5.5 with four equal-Q L-networks. Compare the Q and bandwidth of the two networks.

5-8 For a load and source resistor ratio of 5, find the minimum number of cascaded L sections needed to achieve a loaded Q of 0.5.

5-9. Use the Impedance Matching Utility in ADS to solve a complex load matching example with $R_L = 1000$ Ω in parallel with a 3 pF capacitance. For the source select $R_S = 50$ Ω, and for the load select the parallel RC with R = 1000 Ω and C = 3 pF. Compare the Pi and Tee Matching Network response. Plot the insertion loss of both networks on the same plot.

Chapter 6

Distributed and Microstrip Impedance Matching

6.1 Introduction

Distributed networks are comprised of transmission line elements rather than discrete resistors, inductors, and capacitors. These transmission lines can take the form of the various transmission lines covered in chapter 2. At RF and microwave frequencies, where the wavelengths of the signals become comparable to the physical dimensions of the components, even lumped elements behave like distributed components. At microwave frequencies the distributed networks are more realizable than the lumped element networks.

6.2.2 Analytical Design of Quarter-Wave Matching Networks

In this subsection, based on the equations developed in section 6.2.1, two quarter-wave matching networks for $R_L = 2\ \Omega$ and $R_L = 150\ \Omega$ are designed. For each case the loaded Q factor and bandwidth is calculated at 3 and 20 dB return loss.

Example 6.2-1 Design a quarter-wave network intended to match a 50 Ω source to a 2 Ω load at 100 MHz. Calculate the Q factor and the fractional bandwidths at 3 and 20 dB return loss. Compare the calculations with the simulation results in ADS.

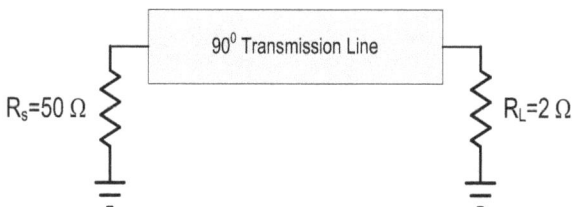

Figure 6-1 Matching quarter-wave transformer ($R_L < R_s$)

Solution: Using the equations developed in the Appendix C, the characteristic impedance, loaded Q factor, and the bandwidths of the matching network are calculated as follows.

1. Enter design parameters and normalize the load impedance

RS=50; RL=2; f=100e6, r=RL/RS

2. Use Equations (6-4) to (6-12) in Appendix C to calculate the parameters

Z1=RS*sqrt(r)

FBW20dB=2-(4/PI)*acos((0.2*sqrt(r))/(sqrt(0.99)*(abs(1-r))))

BW20dB=(f*FBW20dB)

FBW3dB=2-(4/PI)*acos((2*sqrt(r))/(abs(1-r)))

BW3dB=(f*FBW3dB)

Q=1/FBW3dB

The calculation results show that the characteristic impedance = 10 Ω, Q factor is 1.827, and the bandwidths at 3 dB and 20dB return loss are 54.72 MHz and 5.333 MHz, respectively.

The fractional bandwidth at 20 dB return loss indicates that the matching network has a narrow bandwidth. This is characteristic of a matching network where the load resistor is much smaller than the source resistor.

To measure the same parameters in ADS, create a new workspace and open a new schematic window. Insert the S_Params Template and place the TLIN component on the schematic. Set the component values as shown in Figure 6-2.

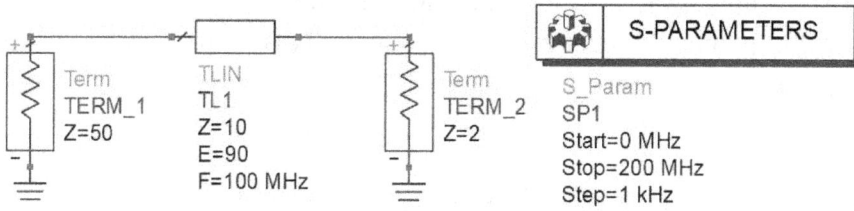

Figure 6-2 Schematic of the quarter-wave matching network

Simulate the schematic and display the input reflection coefficient, S11, and the forward transmission, S21, in a rectangular plot. To measure the 3 dB bandwidth automatically in ADS, use the built-in equation for BW3dB, as shown in Figure 6-3.

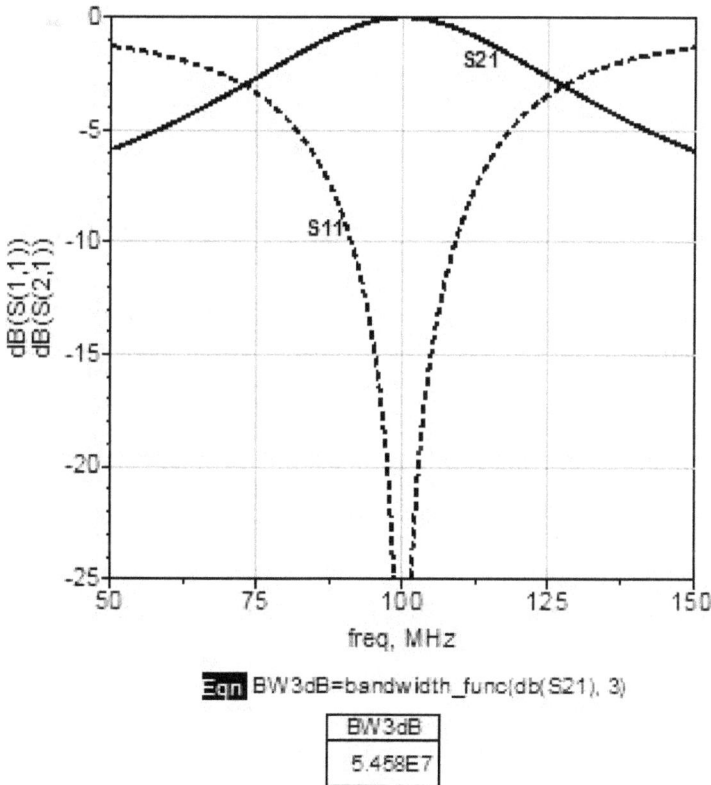

BW3dB=bandwidth_func(db(S21), 3)

BW3dB
5.458E7

Figure 6-3 Network Response and 3 dB bandwidth

Figure 6-3 shows that the 3 dB bandwidth is 54.58 MHz. To measure the bandwidths manually, place markers at 3 and 20 dB return loss as shown in Figure 6-4.

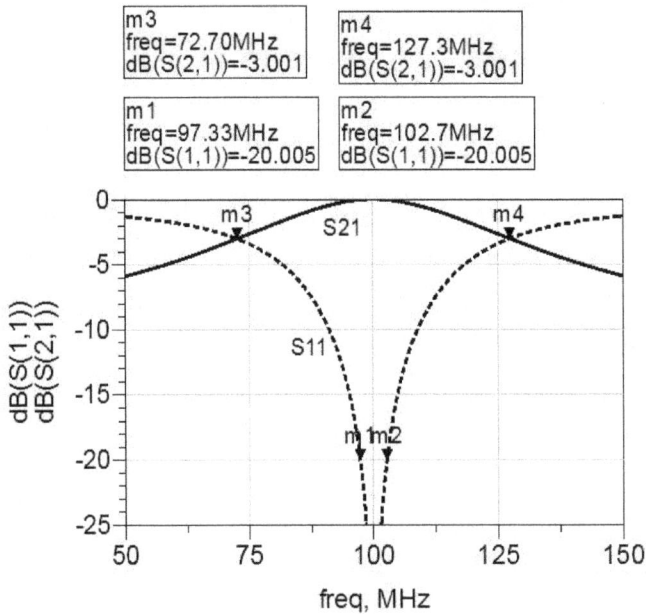

Figure 6-4 Bandwidth measurement at 3 and 20 dB return loss

Reading marker frequencies at 3 and 20 dB return loss, the corresponding bandwidths are:

$$BW_{-3dB} = 127.3 - 72.7 = 54.6 \qquad MHz$$

$$BW_{-20dB} = 102.7 - 97.3 = 5.4 \qquad MHz$$

The Q factor of the matching network is measured by using the equation.

$$Q = \frac{1}{FBW_{-3dB}} \quad \frac{\sqrt{(127.3)(72.7)}}{127.3 - 72.7} = 1.76$$

Note that the measured Q factor and bandwidths are in close agreement with the calculated values.

Example 6.2-2 Design a quarter-wave network intended to match a 50 Ω source to a 150 Ω load. The design frequency is 100 MHz. Compare the calculated Q factor and the fractional bandwidths, at 3 and 20 dB return loss, with the measurements.

Solution: The schematic of the quarter-wave matching network with Rs = 50 and R_L = 150 Ohm is shown in Figure 6-5.

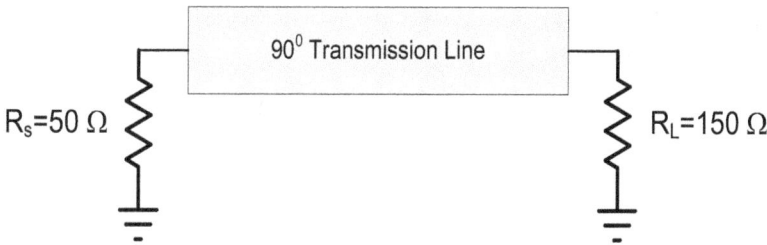

Figure 6-5 Quarter-wave matching network (R_L > R_S)

The characteristic impedance of the quarter-wave matching network is:

$$Z_1 = \sqrt{(50)(150)} = 86.6 \ \Omega$$

The schematic of the matching network in ADS is shown in Figure 6-6.

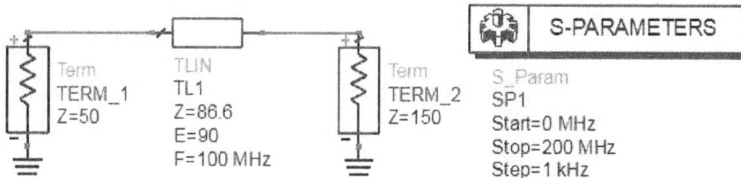

Figure 6-6 Schematic of the matching quarter-wave network

Use equations developed in the Appendix C, calculate the characteristic impedance, the Q factor, and the bandwidths of the matching network at 3 and 20 dB return loss. The procedure follows.

1. Enter design parameters and normalize the load impedance

RL=150; f=100e6, r=RL/RS

2. Use Equations (6-4) to (6-12) in Appendix C to calculate the parameters

Z1=RS*sqrt(r)

FBW20dB=2-(4/PI)*acos((0.2*sqrt(r))/(sqrt(0.99)*(abs(1-r))))

BW20dB=(f*FBW20dB)

FBW3dB=2-(4/PI)*acos((2*sqrt(r))/(abs(1-r)))

BW3dB=(f*FBW3dB)

Qe=1/FBW3dB

The calculation results show that the bandwidth at 20 dB return loss is 22.28 MHz. This is over four times the bandwidth that was achieved with the 2 Ω load impedance. Also notice that the overall Q of the network has significantly increased due to higher load impedance.

Next simulate the schematic and display the input reflection coefficient, S11, and forward transmission, S21, in a rectangular plot. Place markers to measure the bandwidths at 20 dB return loss.

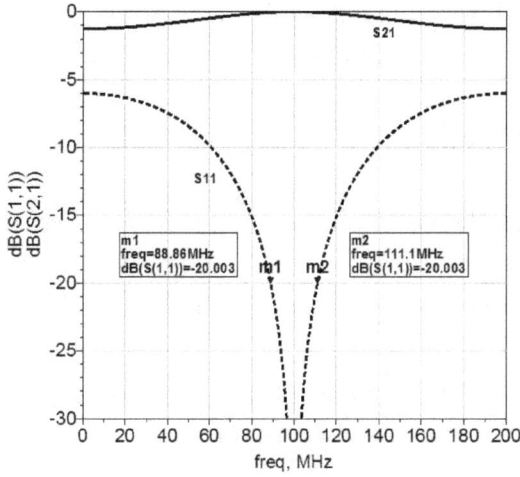

Figure 6-7 Response of the quarter-wave matching network

As Figure 6-7 shows the measured bandwidth at 20 dB return loss is:

$$111.1 - 88.86 = 22.24 \text{ MHz}$$

Quarter-Wave Matching Bandwidth

We have shown that the achievable bandwidth in a quarter-wave matching network is related to the ratio of the load to the source impedance (mismatch ratio) as well as the value of the input reflection coefficient. It is insightful to examine the relationship between these quantities when one of the parameters is swept in value.

Quarter-Wave Matching Bandwidth and Power Loss

The fractional bandwidth and power loss of a quarter-wave matching network can be calculated as a function of the input reflection coefficient, Γ_{IN}, and the normalized load resistor.

The procedure follows.

1. Enter the equations for the input reflection coefficient, the power loss, and the conversion of reflection coefficient to return loss in dB.

2. Enter the desired values for the source and load resistors

3. Normalize the load resistor with respect to source resistor

4. Use Equation (6-6) in Appendix C to calculate the fractional bandwidth.

Example 6.3-1: For a 50 Ω source and 2 Ω load resistors, calculate the fractional bandwidth and power loss from $\Gamma = 0.1$ to $\Gamma = 0.707$.

Solution: Follow the above procedure to calculate the fractional bandwidth and power loss when $R_S = 50$ Ω and $R_L = 2$ Ω.

1. Write the reflection coefficient, power loss, and return loss equations

Vswr = Sweep1_Data.VSWR1

ReflCoef = (vswr-1)/(vswr+1)

Powerloss = (1-(1-(abs(ReflCoef)^2)))*100

RLdB = -20*log(abs(ReflCoef))

1. Enter the source and load resistor values

RS=50; RL=2, r=RL/RS

3. Calculate the Fractional Bandwidth

FBW=(2-(4/PI)*acos((2*(ReflCoef)*sqrt(r))/(sqrt((1-(ReflCoef)^2))*abs(1-r))))

Sweep the parameters to display the fractional bandwidth and power loss for reflection coefficients from 0.1 to 0.707, as shown in Table 6-1.

	ReflCoef	FBW ...	RLdB	Powerloss
1	0.1	0.053	20	1
2	0.11	0.059	19.172	1.21
3	0.12	0.064	18.417	1.44
4	0.13	0.07	17.721	1.69
5	0.14	0.075	17.077	1.96
6	0.15	0.081	16.478	2.25
7	0.16	0.086	15.917	2.56
8	0.17	0.092	15.391	2.89
9	0.18	0.097	14.895	3.24
10	0.19	0.103	14.425	3.61
11	0.2	0.108	13.979	4
12	0.3	0.167	10.458	9
13	0.4	0.233	7.959	16
14	0.5	0.309	6.021	25
15	0.6	0.405	4.437	36
16	0.707	0.547	3.012	49.985

Table 6-1 Fractional bandwidth and power loss for $R_L = 2\ \Omega$

Table 6-1 shows that the fractional bandwidths at 0.1 reflection coefficient, corresponding to 20 dB return loss, is 5.3 % with 1% power loss while at 0.707 reflection coefficient, corresponding to 3 dB return loss, the fractional bandwidth is 54.7 % with 49.985 % power loss. Notice that the fractional bandwidths, at 3 and 20 dB return loss, are the same as calculated in Example 6.2-1.

Example 6.3-2: For a 50 Ω source and 150 Ω load resistors, calculate the fractional bandwidth and power loss from $\Gamma = 0.1$ to $\Gamma = 0.707$.

Solution: Following the procedure, the calculations are shown in Figure 6-15. Sweep the parameters and display the result, as shown in Table 6-3.

1. Enter reflection coefficient, powerloss, and return loss Equations

vswr=Sweep1_Data.VSWR1

ReflCoef=(vswr-1)/(vswr+1)

Powerloss=(1-(1-(abs(ReflCoef)^2)))*100

RLdB=-20*log(abs(ReflCoef))

2. Enter the source and load resistor values and normalized load

RS=50; RL=150; r=RL/RS

2.   Calculate the Fractional Bandwidth

FBW=(2-(4/PI)*acos((2*(ReflCoef)*sqrt(r))/(sqrt((1-(ReflCoef)^2))*abs(1-r))))

The values of the reflection coefficient, power loss, and return loss are given in the Table.6-2.

	ReflCoef	FBW (rad)	RLdB	Powerloss
1	0.1	0.223	20	1
2	0.11	0.246	19.172	1.21
3	0.12	0.269	18.417	1.44
4	0.13	0.292	17.721	1.69
5	0.14	0.315	17.077	1.96
6	0.15	0.339	16.478	2.25
7	0.16	0.362	15.917	2.56
8	0.17	0.386	15.391	2.89
9	0.18	0.411	14.895	3.24
10	0.19	0.435	14.425	3.61
11	0.2	0.46	13.979	4
12	0.3	0.733	10.458	9
13	0.4	1.091	7.959	16
14	0.5	2	6.021	25
15	0.6	1.#QO	4.437	36
16	0.707	1.#QO	3.012	49.985

Table 6-2 Fractional bandwidth and power loss for R_s=50 Ω and R_L=150 Ω

Table 6-2 shows that the fractional bandwidth at $\Gamma = 0.1$, corresponding to 20 dB return loss, is 22.3 % with a 1% power loss.

Notice that the fractional bandwidth is the same as calculated in the previous example. The higher fractional bandwidth in this example, compared to its value in the previous example is due to the lower ratio of the load to source resistance.

Single-Stub Matching Network Design

Example 6.4-1: Design a open-circuited single-stub network to match a load resistance $Z_L = 2 - j5$ Ω to a resistive source $R_S = 50$ Ω at 100 MHz. Calculate the fractional bandwidths of the matching network at 3 and 20 dB return loss. Display the response and compare the calculations with measurements.

Solution: The design example is shown in Figure 6-8.

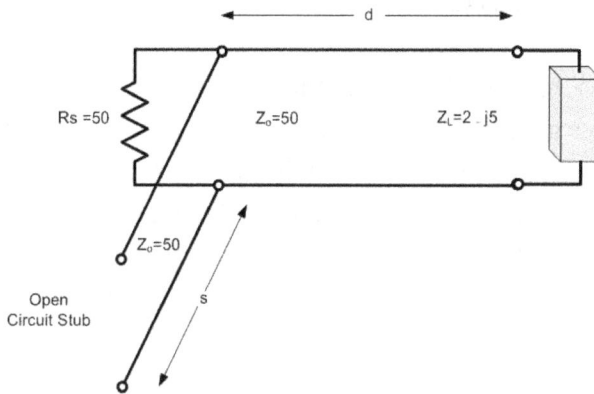

Figure 6-8 Complex load to resistive source matching network

Using Equations , calculate the electrical lengths of the line and stub matching transmission lines. The calculation in Matlab script follows.

1. Enter design parameters and normalized load impedance

RS=50; RL=2; XL=-5; f=100e6; r=RL/RS; x=XL/RS

2.Use Equations (6-18) to (6-29 in Appendix C to calculate t1, t2, d1 and d2, B1, B2, so1 and so2.

t1=(x+sqrt(r*(r^2+x^2-2*r+1)))/(r-1)

t2=(x-sqrt(r*(r^2+x^2-2*r+1)))/(r-1)

d1=360*(atan(t1))/(2*pi)

d2=360*(atan(t2))/(2*pi)

B1=(x*t1^2+(r^2+x^2-1)*t1-x)/(RS*(r^2+x^2+t1^2+2*x*t1))

B2=(x*t2^2+(r^2+x^2-1)*t2-x)/(RS*(r^2+x^2+t2^2+2*x*t2))

so1=360*(pi-atan(RS*B1))/(2*pi)
so2=-360*atan(RS*B2)/(2*pi)

The calculation results show that the problem has two solutions. Either solution can be used in the design of the single-stub matching network. However, for both line and stub, usually the shorter line lengths are chosen. For this example we select the shorter electrical lengths in the second solution; d2 for the line and so2 for the open-circuited stub. The schematic of the single-stub matching network is shown in Figure 6-9.

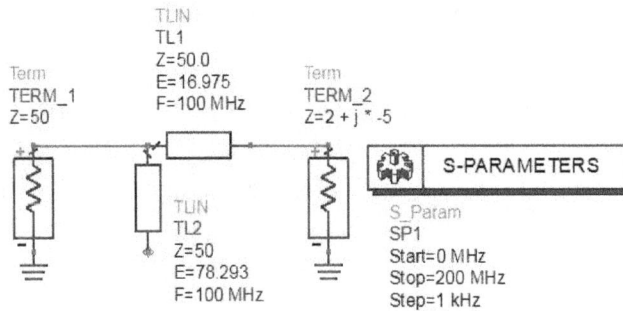

Figure 6-9 Schematic of the single-stub matching network

Simulate the schematic and display S11 and S21 in a rectangular plot.

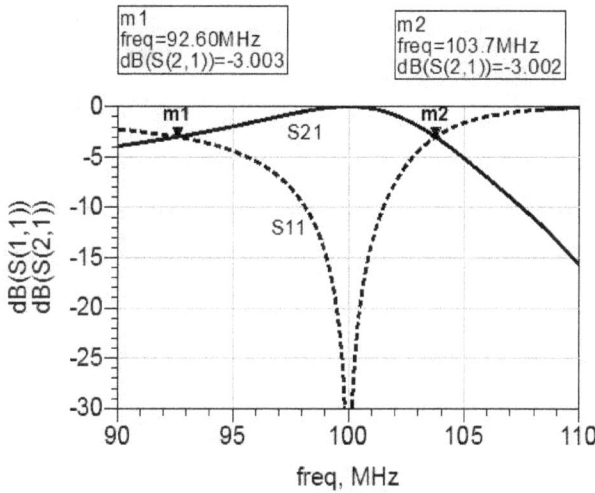

Figure 6-10 Response of the single-stub matching network

Notice that the matching bandwidth at 3 dB return loss is about 11% while at 20 dB is only 1%. This suggests a very narrow matching bandwidth.

Graphical Design of Single-Stub Matching Networks

In this section a 50 Ω source will be matched to a 5 - 25j Ω load at a frequency of 1000 MHz. Both open-circuited and short-circuited shunt stubs will be considered.

Example 6.5-1 Design a single-stub network to match the 5 - j25 Ω load impedance to 50 Ωsource resistor. Use an open-circuited stub.

Solution: Create a new workspace in ADS and open a new schematic window. Insert the S_Params Template and add a series transmission line (TLIN) between the source and load terminations. Wire up the components and set the component values as shown in Figure 6-11.

Because we intend to place a shunt element after the series transmission line, enable the admittance circles on the Smith Chart. Make the length of the line tunable and tune it to move the load impedance (point A) to intersect the unit conductance circle on the top half of the Smith chart (point B. As the schematic shows, an electrical length of 42.5° places the impedance on the unit conductance circle at point B.

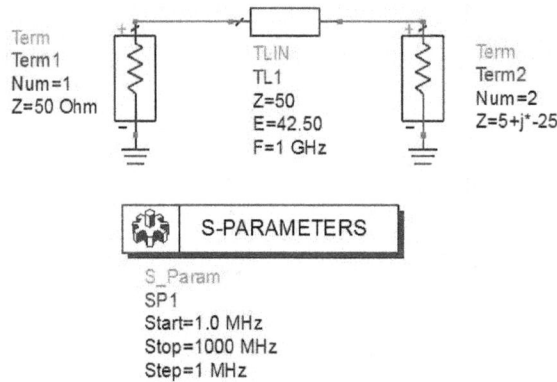

Figure 6-11 Adding series transmission line (electrical length=42.5°)

The movement of the impedance from point A to point B is displayed in Figure 6-12.

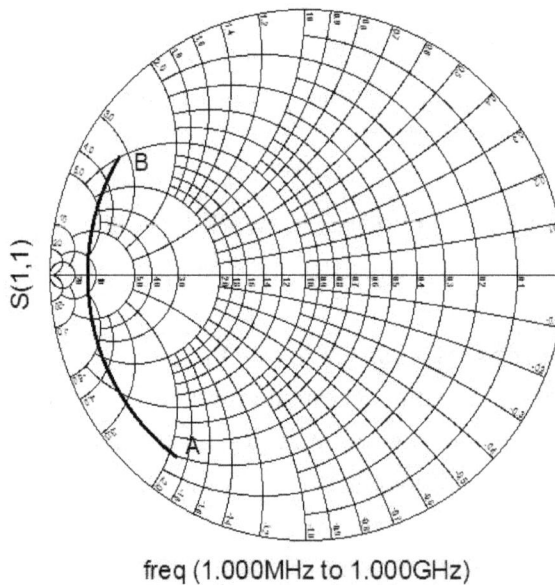

freq (1.000MHz to 1.000GHz)

Figure 6-12 Point B on the unit conductance circle (electrical length=42.5°)

At point B the normalized impedance is z = 0.088 + 0.28j Ω. To move point B to the center of Smith chart, add an open-circuited shunt transmission line and tune the length of the line until the impedance moves to the center of the chart (50Ω). As the schematic of Figure 6-13 shows, a 73° length of transmission line would be required.

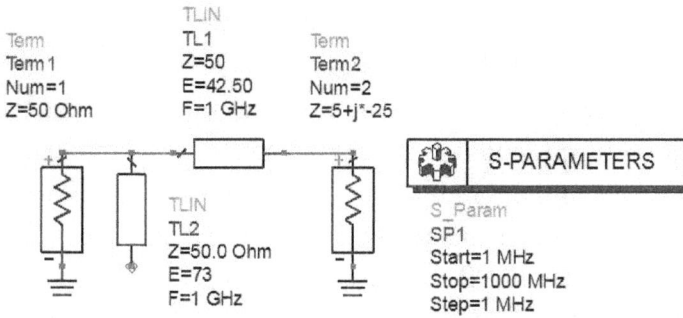

TLIN
TL1
Z=50
E=42.50
F=1 GHz

Term
Term 1
Num=1
Z=50 Ohm

Term
Term2
Num=2
Z=5+j*-25

S-PARAMETERS

TLIN
TL2
Z=50.0 Ohm
E=73
F=1 GHz

S_Param
SP1
Start=1 MHz
Stop=1000 MHz
Step=1 MHz

Figure 6-13 Adding open-circuited shunt transmission line (73°) to network

The movement of point B to the center of the Smith chart is shown in Figure 6-14.

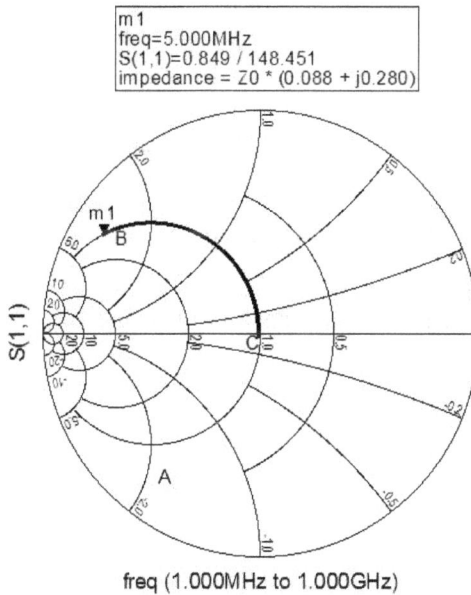

m1
freq=5.000MHz
S(1,1)=0.849 / 148.451
impedance = Z0 * (0.088 + j0.280)

S(1,1)

freq (1.000MHz to 1.000GHz)

Figure 6-14 Moving point B to the center of Smith chart (73°)

6.5.2 Graphical Design Using Short-Circuited Stubs

In this Section a short-circuited shunt stub will be used.

Example 6.5-2 Consider matching the 5 - j25 Ω load impedance to 50 Ω source resistor using short-circuited shunt stubs.

Solution: A short-circuited shunt transmission line can be used to perform the function of the shunt stub. Using a short-circuited shunt transmission line, the series transmission line should intersect the unit conductance circle on the bottom half of the Smith Chart. Add a series transmission line of 11° electrical length to move the impedance at point A to intersect the unit conductance circle at point B as shown in Figure 6-20. Then add the short-circuited shunt transmission line as shown in Figure 6-15. Tune the length to 17° to move the impedance to the center of the Smith Chart.

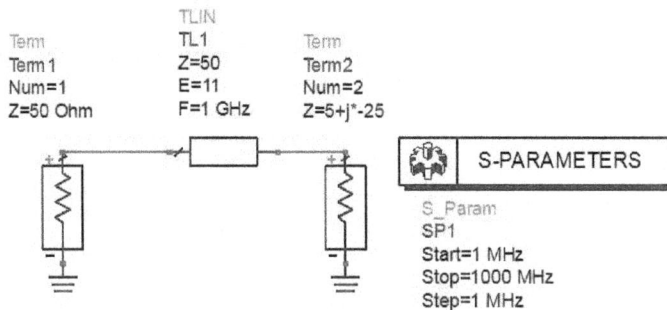

Figure 6-15 Adding series transmission line (11°) to 5-j25 Ω

Simulate the schematic and notice the movement of the impedance from point A to point B.

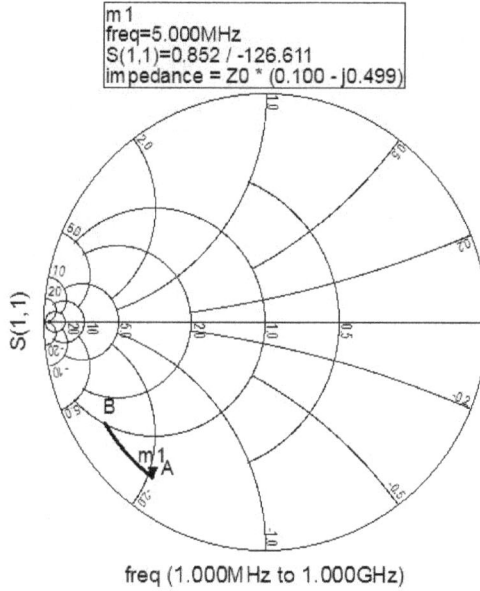

m 1
freq=5.000MHz
S(1,1)=0.852 / -126.611
impedance = Z0 * (0.100 - j0.499)

freq (1.000MHz to 1.000GHz)

Figure 6-16 Adding series transmission line (11°) to 5-j25 Ω

To move point B to the center of Smith chart, add a short-circuited shunt transmission line and tune the length of the line until the impedance moves to the center of the chart (50Ω). As the schematic of Figure 6-17 shows, a 17° length of transmission line would be required.

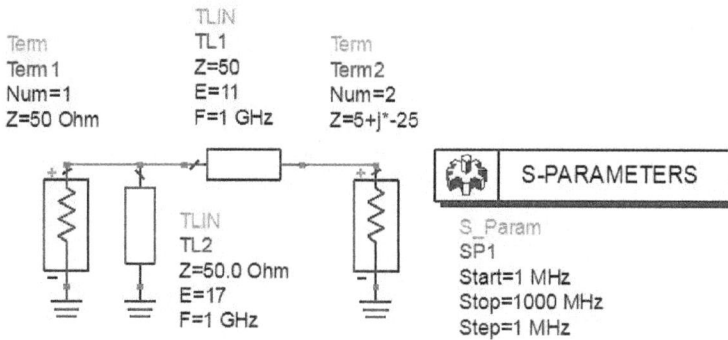

Figure 6-17 Adding short-circuited shunt transmission line (17°)

Simulate the schematic and notice the movement of the impedance from point B to the center of Smith chart.

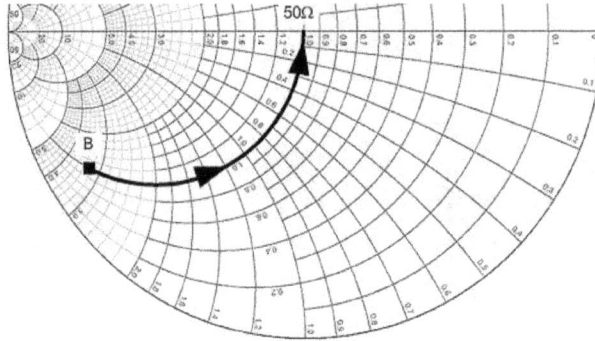

Figure 6-18 Adding short-circuited shunt transmission line (17°)

Design of the Cascaded Matching Networks

When the impedance of the load and the source are both complex functions, we can define a virtual resistor and design single-stub networks to match the complex impedances to the virtual resistor. The final matching network is obtained by cascading the single-stub matching networks. The procedure is demonstrated in the following example.

Example 6.6-1: Design single-stub networks to match a complex load $Z_L = 10 - j5$ Ω to a complex source $Z_S = 50 - j15$ Ω at 100 MHz. Calculate the electrical lengths of the lines and the fractional bandwidths of the matching network at 3 dB and 20 dB return loss.

The procedure follows:

1. Find the intermediate resistor,

$$R_1 = \sqrt{R_L R_S} = \sqrt{(10)(50)} = 22.36 \ \Omega$$

2. Design a single-stub matching network between the intermediate resistor and the load impedance

3. Design a single-stub matching network between the intermediate resistor and the source impedance

4. Cascade the two matching networks

First Part Solution: Design the single-stub matching circuit network between the load impedance and the intermediate resistor

1. Enter design parameters with RS as the intermediate resistor

RS=22.36; RL=50; XL=-15; f=100e6, r=RL/RS, x=XL/RS

2. Use Equations (6-18) to (6-29) in Appendix C to calculate t1, t2, d1 and d2, B1, B2, so1 and so2.

t1=(x+sqrt(r*(r^2+x^2-2*r+1)))/(r-1)

t2=(x-sqrt(r*(r^2+x^2-2*r+1)))/(r-1)

d1=360*(atan(t1))/(2*pi)

d2=360*(pi+atan(t2))/(2*pi)

B1=(x*t1^2+(r^2+x^2-1)*t1-x)/(RS*(r^2+x^2+t1^2+2*x*t1))

B2=(x*t2^2+(r^2+x^2-1)*t2-x)/(RS*(r^2+x^2+t2^2+2*x*t2))

so1=360*(pi-atan(RS*B1))/(2*pi)

so2=-360*atan(RS*B2)/(2*pi)

From calculation results we select d2=114.019 and so2=43.245.

Design of the first matching network is shown in Figure 6-19.

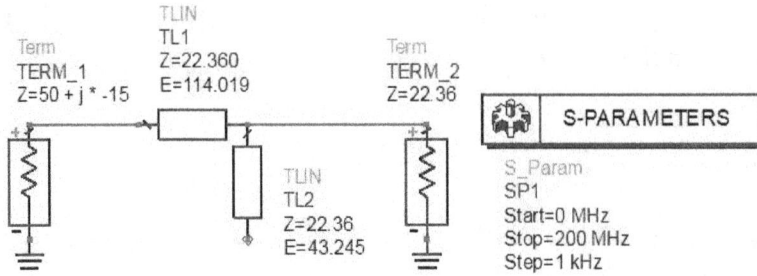

Figure 6-19 Schematic of the first matching network

Second Part Solution: Design the single-stub matching network between the source impedance and the intermediate resistor

1. Enter design parameters and normalize the load impedance

RS=22.36; RL=10; XL=-5; f=100e6, r=RL/RS' x=XL/RS

2. Use Equations (6-18) to (6-29) in Appendix C to calculate t1, t2, d1 and d2, B1, B2, so1 and so2.

t1=(x+sqrt(r*(r^2+x^2-2*r+1)))/(r-1)

t2=(x-sqrt(r*(r^2+x^2-2*r+1)))/(r-1)
d1=360*(pi+atan(t1))/(2*pi)

d2=360*(atan(t2))/(2*pi)

B1=(x*t1^2+(r^2+x^2-1)*t1-x)/(RS*(r^2+x^2+t1^2+2*x*t1))

B2=(x*t2^2+(r^2+x^2-1)*t2-x)/(RS*(r^2+x^2+t2^2+2*x*t2))

so1=360*(pi-atan(RS*B1))/(2*pi)

so2=-360*atan(RS*B2)/(2*pi)

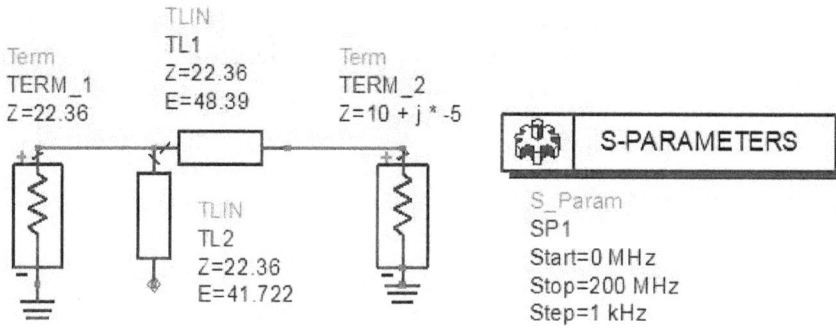

Figure 6-20 Schematic of the second matching network

Now cascade both line and stub matching networks to obtain the complete matching network, as shown in Figure 6-21.

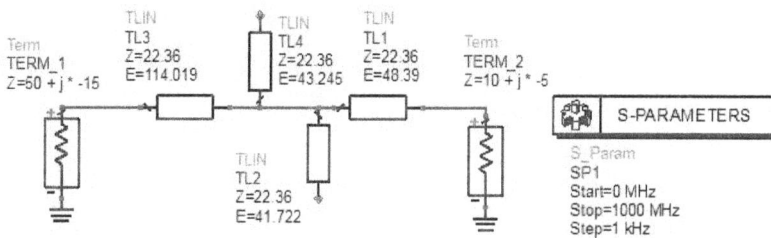

Figure 6-21 Cascading single-stub matching networks

The simulated response of the complete matching network is shown in Figure 6-22. Notice that the matching network has about 100 % fractional bandwidth at 3 dB return loss and 47% fractional bandwidth at 12 dB return loss. The measured 3 dB bandwidth between the markers is exactly the same as calculated by the equations.

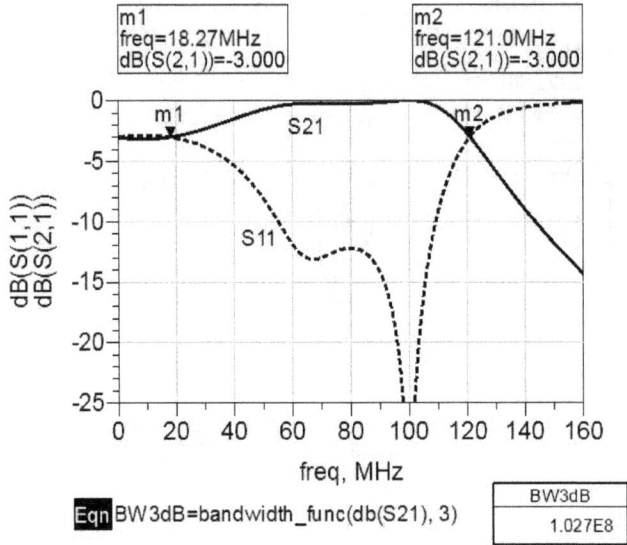

m1
freq=18.27MHz
dB(S(2,1))=-3.000

m2
freq=121.0MHz
dB(S(2,1))=-3.000

Eqn BW3dB=bandwidth_func(db(S21), 3)

BW3dB
1.027E8

Figure 6-22 Response of the cascaded matching network

Broadband Design of the Quarter-Wave Matching Network

In Examples 6.2-1 the bandwidth of a single quarter-wave transformer matching network is less than 10 % which is considered to be a narrow-band matching network. We can increase the bandwidth by cascading two or more quarter-wave transformers to achieve a broadband matching network. To analytically design a broadband matching network with N quarter-wave transformers first use Equation (6-32) to calculate the characteristic impedance of each quarter-wave transformer then cascade all the sections into one matching network. Let Rs and R_L be the source and load impedances to be matched by the N quarter-wave transformation network. The characteristic impedance of each section can be calculated from the following equation.

$$Z_n = R_S(r)^{(2n-1)/2N} \qquad n = 1, 2, \dots, N$$

Where $r = R_L/R_S$ is the normalized load resistance and N is the number of quarter-wave transformers.

Example, 6.7-1 Design a three-section quarter-wave transformer network to match a load resistance $R_L = 2\,\Omega$ to a source $R_S = 50\,\Omega$ at 100 MHz.

(a) Calculate the characteristic impedance of the quarter-wave transformers, the Q factor and the fractional bandwidths of the matching network at 3 and 20 dB return loss.

(b) Display the simulated response and compare the measurements with the measurements of single quarter-wave transformer matching network.

Solution: (a) Using the above equation for Z_n the characteristic impedances of the 3-section quarter-wave transformers are:

$$Z1 = R_S(r)^{1/2N} = 50(0.04)^{1/6} = 29.24$$

$$Z2 = R_S(r)^{3/2N} = 50(0.04)^{3/6} = 10.00$$

$$Z3 = R_S(r)^{5/2N} = 50(0.04)^{5/6} = 3.42$$

The intermediate resistors can be obtained from equation below

$$R_n = R_S(r)^{\frac{n}{N}} \quad n = 1, 2, 3, \dots, N\text{-}1$$

Therefore, the two intermediate resistors are:

$$R_1 = 50\,(0.04)^{\frac{1}{3}} = 17.1\,\Omega$$

$$R_2 = 50\,(0.04)^{\frac{2}{3}} = 5.848\,\Omega$$

The schematic of the three-section quarter-wave matching network is shown in Figure 6-23.

1. Enter design parameters and normalize the load impedance

RS=50; RL=2; f=100e6, r=RL/RS, N=3

3. Calculate characteristic impedances and intermediate impedances

Z1=RS*r^(1/(2*N))

Z2=RS*r^(3/(2*N))

Z3=RS*r^(5/(2*N))

R1=RS*r^(1/N); R2=RS*r^(2/N)

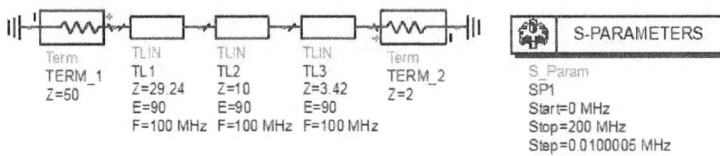

Figure 6-23 Three-section quarter-wave matching network

The simulated response of the matching network is shown in Figure 6-24.

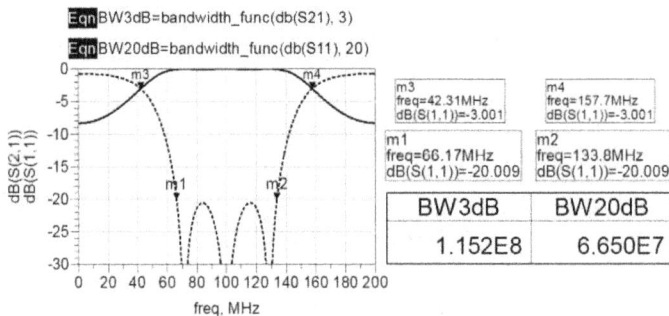

Figure 6-24 Response of the cascaded matching network

Figure 6-24 shows that the simulated bandwidth at the 3 dB return loss is:

$$BW_{3dB} = 157.7 - 42.3 = 115.4 \text{ MHz}$$

$$\text{FBW}_{3dB} = \frac{115.4 x 100}{\sqrt{(157.7).(42.3)}} = 140.8\%$$

Similarly the simulated bandwidth at the 20 dB return loss is:

$$\text{BW}_{20dB} = 133.8 - 66.2 = 67.6 \text{ MHz}$$

$$\text{FBW}_{20dB} = \frac{67.6 x 100}{\sqrt{(133.8).(66.2)}} = 71.8\%$$

The measured Q factor for the matching network is,

$$Q = \frac{1}{1.408} = 0.71$$

The wide bandwidth and lower Q factor is an indication that by adding two more quarter-wave sections the Q factor has reduced to less than one half and the 3-dB bandwidth has more than doubled compared to a single quarter-wave matching network.

Example 6.7-2: Design a broadband quarter-wave network to match a complex load, $Z_L = 10 - j5 \ \Omega$ to 50 Ω source impedance at 100 MHz. Calculate the bandwidth at 20 dB return loss and compare with the bandwidth of a single-stub matching network.

Solution: To design a broadband matching network between a resistive source and complex load impedance, first we design a 5 section matching network between the real source and the real part of the load impedance. Then we replace the quarter-wave transformer adjacent to the load with a single-stub network that matches the complex load to the real resistor. The procedure follows:

1. Enter design parameters and normalize the load impedance
   RS=50; RL=10;f=100e6; r=RL/RS; N=5

2. Calculate characteristic impedances and intermediate resistors

Z1=RS*r^(1/(2*N)) = 42.567

Z2=RS*r^(3/(2*N)) = 30.852

Z3=RS*r^(5/(2*N)) = 22.361

Z4=RS*r^(7/(2*N)) = 16.207

Z5=RS*r^(9/(2*N)) = 11.746

R1=RS*r^(1/N) = 36.239

R2=RS*r^(2/N) = 26.265

R3=RS*r^(3/N) = 19.037

R4=RS*r^(4/N) = 13.797

The calculation results show that the matching network is:

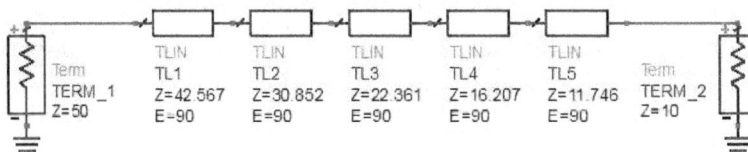

Figure 6-25 Five-section quarter-wave transformer matching network

Next replace the quarter-wave section adjacent to the load with a single-stub matching network between R4 = 13,795 Ohm and Z_L = 10 - j5 Ohm. Based on following calculations, the single-stub matching network is shown in Figure 6-26.

1. Enter design parameters and normalize the load impedance

RS=13.797; RL=10; XL=-5; f=100e6, r=RL/RS, x=XL/RS 2.

2. Use the equations in Appendix C to calculate line and stub values

t1=(x+sqrt(r*(r^2+x^2-2*r+1)))/(r-1)

t2=(x-sqrt(r*(r^2+x^2-2*r+1)))/(r-1)

d1=360*(pi+atan(t1))/(2*pi)

d2=360*(atan(t2))/(2*pi)

B1=(x*t1^2+(r^2+x^2-1)*t1-x)/(RS*(r^2+x^2+t1^2+2*x*t1))

B2=(x*t2^2+(r^2+x^2-1)*t2-x)/(RS*(r^2+x^2+t2^2+2*x*t2))

so1=360*(pi-atan(RS*B1))/(2*pi)

so2=-360*atan(RS*B2)/(2*pi)

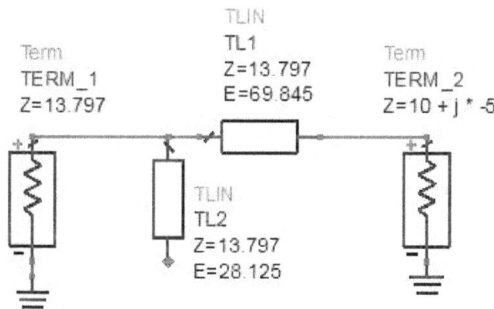

Figure 6-26 Schematic of the single-stub matching network

Now cascade the line and stub with four quarter-wave networks to form the final design of the matching network as shown in Figures 6-27.

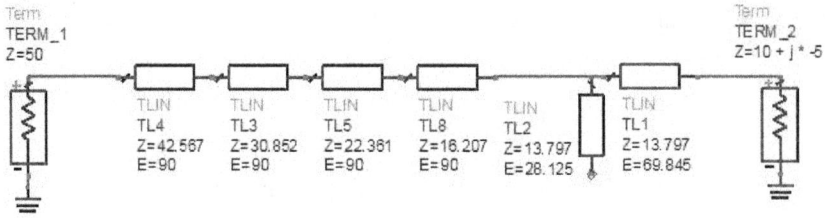

Figure 6-27 Schematic of the broadband matching network

The simulated response of the matching network is shown in Figure 6-28.

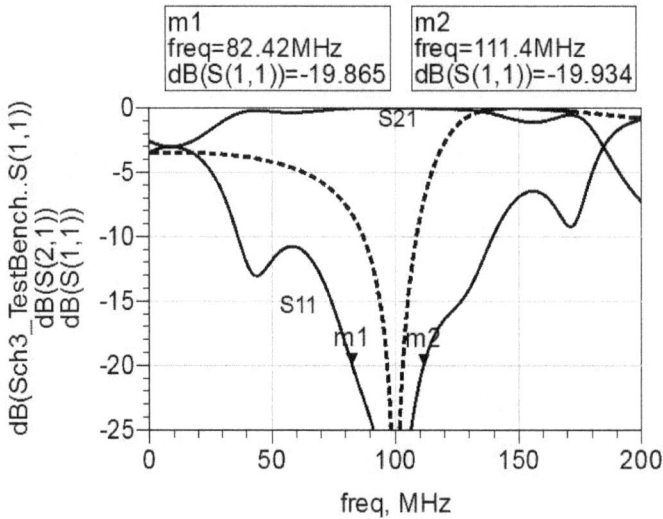

Figure 6-28 Response of the broadband matching network

Figure 6-28 shows that the bandwidth at 20 dB return loss is:

$$BW_{20dB} = 111 - 82 = 29 \text{ MHz}$$

And the fractional bandwidth at 20 dB return loss is:

$$FBW_{20dB} = \frac{29}{\sqrt{(111).(82)}} = 30.3 \text{ \%}$$

The same numbers for the single-stub matching network are:

$$BW_{20dB} = 102.8 - 96.8 = 6 \quad MHz$$

$$FBW_{20dB} = \frac{6x100}{\sqrt{(102.8).(96.8)}} = 6\%$$

Notice that fractional bandwidth for the broadband at 20 dB return loss is 30.3 % as opposed to only 6 % for the narrowband single-stub matching network. Therefore, we have increased the matching bandwidth at 20 dB return loss by five times over the single-stub matching network.

Example 6.7-3: Design a broadband network to match a complex load, $Z_L = 150 - j30$ Ohm to 50 Ω source impedance at 100 MHz. Display the simulated response and measure the bandwidth at 20 dB return loss. Compare the results with the singe-stub matching network.

Solution: Use the same method as in Example 6.6.1 to design the broadband matching network. The design of the broadband quarter-wave transformer matching network with five quarter-wave sections is shown in Figure 6-29.

1. Enter design parameters and normalize the load impedance

RS=50; RL=150; f=100e6, r=RL/RS, N=5

2. Calculate characteristic impedances and intermediate resistors

Z1=RS*r^(1/(2*N)) = 55.806

Z2=RS*r^(3/(2*N)) = 69.915

Z3=RS*r^(5/(2*N)) = 86.603

Z4=RS*r^(7/(2*N)) = 107.883

Z5=RS*r^(9/(2*N)) = 134.394

R1=RS*r^(1/N) = 62.287

R2=RS*r^(2/N)=77.592   =

R3=RS*r^(3/N) = 96.659

R4=RS*r^(4/N) = 120.411

The ADS schematic of the five-section quarter-wave matching network is shown in Figure 6-29. The simulated response of the quarter-wave matching network is shown in Figure 6-30.

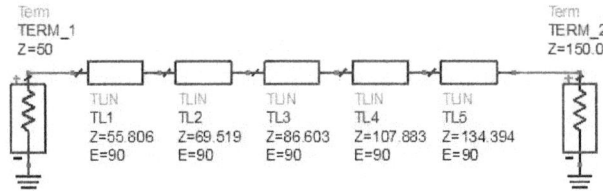

Figure 6-29 Five-section quarter-wave matching network

The wideband response is shown in Figure 6-30.

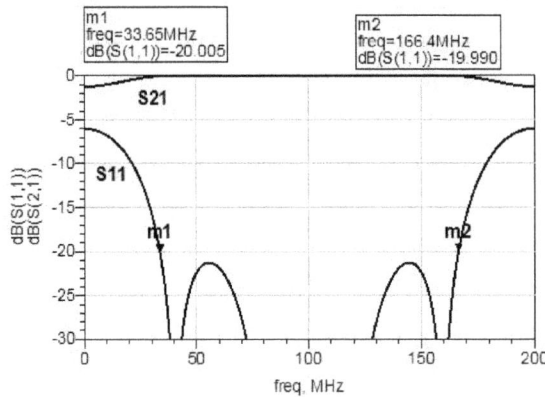

Figure 6-30 Response of the matching network

Next replace the quarter-wave transformer adjacent to the load with a single-stub matching network. The design of the single-stub matching network is shown in the following Equation Editor. Note that the source resistor, R4 = 13.797 Ω, was calculated.

1. Enter design parameters and normalize the load impedance

RS=120.411; RL=150; XL=-30; f=100e6, r=RL/RS, x=XL/RS

2. Use the equations in Appendix C to calculate t1 and t2, d1 and d2,

B1 and B2, so1 and so2

t1=(x+sqrt(r*(r^2+x^2-2*r+1)))/(r-1)

t2=(x-sqrt(r*(r^2+x^2-2*r+1)))/(r-1)

d1=360*(atan(t1))/(2*pi)

d2=360*(pi+atan(t2))/(2*pi)

B1=(x*t1^2+(r^2+x^2-1)*t1-x)/(RS*(r^2+x^2+t1^2+2*x*t1))

B2=(x*t2^2+(r^2+x^2-1)*t2-x)/(RS*(r^2+x^2+t2^2+2*x*t2))

so1d=360*(pi-atan(RS*B1))/(2*pi)

so2d=-360*atan(RS*B2)/(2*pi)

The single-stub matching network is shown in Figure 6-31.

TLIN
TL1
Z=120.411
E=29.922

Term
TERM_1
Z=120.411

Term
TERM_2
Z=150 + j * -30

TLIN
TL2
Z=120.411
E=162.592

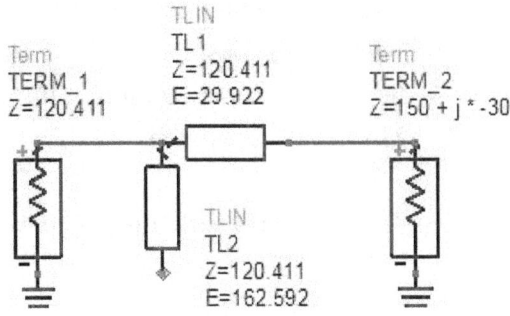

Figure 6-31 Schematic of the single-stub matching network

Now cascade the two matching networks to form the final design of broadband matching network, as shown in Figure 6-32.

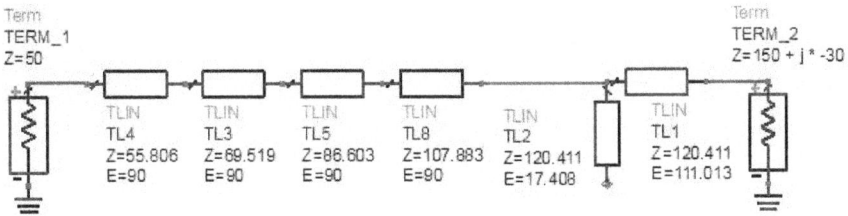

Term
TERM_1
Z=50

Term
TERM_2
Z=150 + j * -30

TLIN
TL4
Z=55.806
E=90

TLIN
TL3
Z=69.519
E=90

TLIN
TL5
Z=86.603
E=90

TLIN
TL8
Z=107.883
E=90

TLIN
TL2
Z=120.411
E=17.408

TLIN
TL1
Z=120.411
E=111.013

Figure 6-32 Schematic of the broadband matching network

The simulated response of the quarter-wave matching network is shown in Figure 6-33.

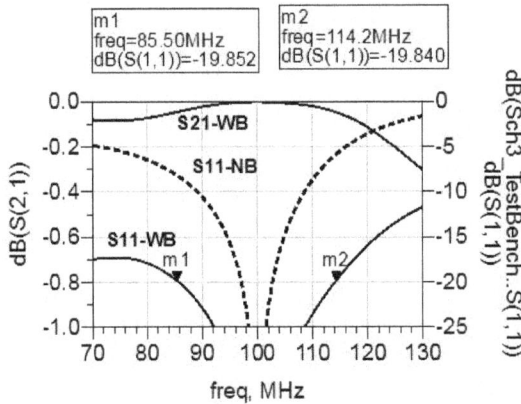

m1
freq=85.50MHz
dB(S(1,1))=-19.852

m2
freq=114.2MHz
dB(S(1,1))=-19.840

Figure 6-33 Response of the broadband matching network

Figure 6-33 shows that the bandwidths of the broadband matching network at 20 dB return loss are:

$$BW_{20dB} = 114.2 - 86.0 = 27.8 \text{ MHz}$$

$$FBW_{20dB} = \frac{27.8 x 100}{\sqrt{(113.8).(86.0)}} = 28.1 \%$$

The same numbers for the narrowband single-stub matching network are:

$$BW_{20dB} = 102.8 - 97.1 = 5.7 \text{ MHz}$$

$$FBW_{20dB} = \frac{5.7 x 100}{\sqrt{(102.8).(97.1)}} = 5.7 \%$$

Notice that the fractional bandwidth at 20 dB return loss is about 28.1 % as opposed to only 5.7 % for the narrowband matching network. This is an indication that, by adding four quarter-wave sections to the single-stub matching network, we have increased the matching bandwidth at 20 dB return loss nearly five times over a single-stub matching network.

References and Further Reading

[1]   Ali A. Behagi, *RF and Microwave Circuit Design,* A Design Approach Using (**ADS**). Techno Search, Ladera Ranch, CA 92694. August 2015.

[2]   Keysight Technologies, Manuals for Advanced Design System, *ADS 2016.0 Documentation Set*, EEsof EDA Division, Santa Rosa, California www.keysight.com

[3]   Guillermo Gonzales, *Microwave Transistor Amplifiers – Analysis and Design*, Second Edition, Prentice Hall Inc., Upper Saddle River, NJ.

[4]   Randy Rhea, *The Yin-Yang of Matching: Part 1 – Basic Matching Concepts*, High Frequency Electronics, March 2006

[5]   Steve C. Cripps, *RF Power Amplifiers for Wireless Communications*, Artech House Publishers, Norwood, MA. 1999.

[6]   David M. Pozar, *Microwave Engineering*, Third Edition, John Wiley & Sons, New York, 2005

[7]   R. Ludwig, P. Bretchko, *RF Circuit Design*, Theory and Applications, Prentice Hall, Upper Saddle River, NJ, 2000

Problems

6-1.   Design a quarter-wave transmission line to match a load resistance $R_L = 5\ \Omega$ to a resistive source $R_S = 75\ \Omega$ at 500 MHz. Calculate the characteristic impedance of the quarter-wave line, the Q factor and the fractional bandwidths of the matching network at 3 and 20 dB return loss. Display the simulated response and compare the calculations with measurements. For the quarter-wave matching network, calculate the fractional bandwidth and power loss from $\Gamma = 0.1$ to $\Gamma = 0.707$.

6-2.   Design a quarter-wave transmission line to match a load resistance $R_L = 100\ \Omega$ to a resistive source $R_S = 25\ \Omega$ at 600 MHz. Calculate the characteristic impedance of the quarter-wave line, the Q factor and the fractional bandwidths of the matching network at 3 dB and 20 dB return loss. Display the simulated response and compare the calculations with measurements. For the quarter-wave matching network, calculate the fractional bandwidth and power loss from $\Gamma = 0.1$ to $\Gamma = 0.707$.

6-3.   Design a single-stub network to match a load resistance $Z_L = 5 - j5$ Ω to a resistive source $R_S = 75$ Ω at 700 MHz. Calculate the electrical lengths of the matching line and stub and the fractional bandwidths of the matching network at 3 and 20 dB return loss. Display the simulated response and compare the calculations with measurements.

6-4.   Design a single-stub network to match a load resistance $Z_L = 100 + j20$ Ω to a resistive source $R_S = 40$ Ω at 800 MHz. Calculate the electrical lengths of the matching line and stub and the fractional bandwidths of the matching network at 3 dB and 20 dB return loss. Display the simulated response and compare the calculations with measurements.

6-5.   Design a single-stub network to match a complex load $Z_L = 20 + j5$ Ω to a complex source $Z_S = 50 + j20$ Ω at 1000 MHz. Calculate the electrical lengths of the matching line and stub and the fractional bandwidths of the matching network at 3 dB and 20 dB return loss. Display the simulated response and verify the calculations with measurements.

6-6.   Design a three-section quarter-wave network to match a load resistance $R_L = 5$ Ω to a resistive source $R_S = 50$ Ω at 100 MHz. Calculate the characteristic impedance of the quarter-wave line, the Q factor and the fractional bandwidths of the matching network at 3 and 20 dB return loss. Display the simulated response and compare the measurements with the singe quarter-wave matching network.

6-7.   Design a three-section quarter-wave network to match a load resistance $R_L = 100$ Ω to a resistive source $R_S = 25$ Ω at 900 MHz. Calculate the fractional bandwidths of the matching network at 3 and 20 dB return loss. Display the simulated response and compare the measurements with the singe quarter-wave matching network.

6-8.   Design a broadband network to match a complex load, $Z_L = 25 + j10$ Ω to 75 Ω source impedance at 300 MHz. Measure the bandwidth at 20 dB return loss and compare it with the results of singe-stub matching network.

6-9.   Design a broadband network to match a complex load, $Z_L = 100 + j20$ to 40 Ω source impedance at 1200 MHz. Display the simulated response and measure the bandwidth at 20 dB return loss. Compare the results with the singe-stub matching network.

Chapter 7

Single Stage Amplifier Design

Introduction

In this chapter we design four single stage amplifiers that require four different impedance matching techniques.

1. Maximum Gain Amplifier Design
2. Specific Gain Amplifier Design
3. Low Noise Figure Amplifier Design
4. Power Amplifier Design

Most amplifier designs actually involve selective mismatching of the transistor to its source and load impedance to accomplish its intended purpose. Only the maximum gain amplifier requires a conjugate matching network design.

Maximum Gain Amplifier Design

This section covers the design of the maximum gain amplifier at 2.35 GHz using the SHF-0189 transistor. The SHF-0189 is a high performance Hetrostructure FET (HFET) housed in a surface-mount plastic package (SOT-89) as shown in Figure 7-1. If the device is potentially unstable in the band of interest, a conjugate impedance match cannot be realized unless we add additional circuitry to stabilize the device.

SHF-0189

SHF-0189Z (Pb) RoHS Compliant & Green Package

0.05 - 6 GHz, 0.5 Watt
GaAs HFET

Product Features
• Now available in Lead Free, RoHS
  Compliant, & Green Packaging
• High Linearity Performance at 1.96 GHz
  +27 dBm P1dB
  +40 dBm Output IP3
  +16.5 dB Gain
• High Drain Efficiency
• See App Note AN-031 for circuit details

Figure 7-1 GaAs HFET specifications (courtesy of RF Micro Devices)

Stabilizing the Device in ADS

The ADS software has built in functions to determine the stability factor and stability measure as well as Γ_{MS} and Γ_{ML} , the conjugate match reflection coefficients.

Example 7.2-1: Measure and display the stability factor, K, the stability measure, B1, and the maximum gain, Gmax, of the SHF-0189 FET transistor from 100 to 8000 MHz. Design the stability network if the transistor is unstable.

Solution: Create a new workspace in ADS and open a new schematic window. Insert the S_Params Template and place a two port S parameter file, S2P, from Data Items palette on the schematic. Ground the reference port and wire up the schematic. Double click on S2P icon to open the 2-Port S-parameter File dialog window. Browse the File Name and select the SHF-0189.s2p file from any folder in your computer. Set the component values as shown in Figure 7-2.

Ex7_2_1A_lib_TWO_N_FET
Q1
FILENAME="C:\S_Parameters\SHF-0189.s2p"

S-PARAMETERS

Term
TERM_1

Term
TERM_2

S_Param
SP1
Start=100 MHz
Stop=8000 MHz
Step=1.975 MHz

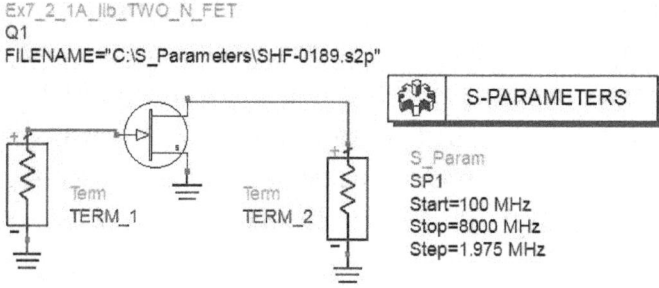

Figure 7-2 Schematic of the SHF-0189 device

Sweep the frequency range of the device over the entire range of frequencies contained in the S parameter file. When the Data Display window opens, write 3 equations to measure the stability factor, K, the stability measure, B1, and the maximum gain, Gmax, as follows.

$$K=stab\_fact(S),$$
$$B1=stab\_meas(S),$$
$$Gmax=max\_gain(S)$$

Plot the parameters K, B1, and Gmax for the SHF-0189 device as shown in Figure 7-3.

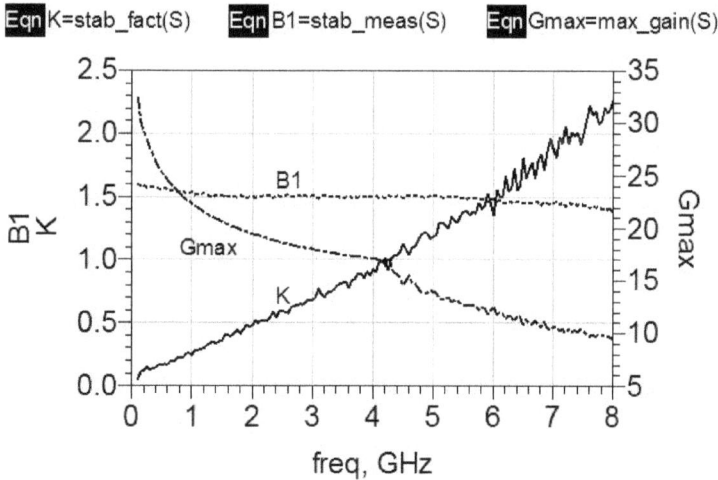

Eqn K=stab_fact(S)   Eqn B1=stab_meas(S)   Eqn Gmax=max_gain(S)

Figure 7-3 Stability parameters and GMAX for the SHF-0189 device

Note that stability factor K is less than 1 from 100 MHz up to about 4500 MHz indicating that the transistor is potentially unstable over most of its frequency range including the design frequency of 2350 MHz. An effective stabilization network for medium and high power transistors employs a parallel RC circuit on the input of the device. Make both the R and C values tunable with a starting value of 10 Ω and 10 pF respectively. Tune the resistance and capacitance values until K > 1 for frequencies above 500 MHz. The values shown in Figure 7-4 provide sufficient stability above 500 MHz.

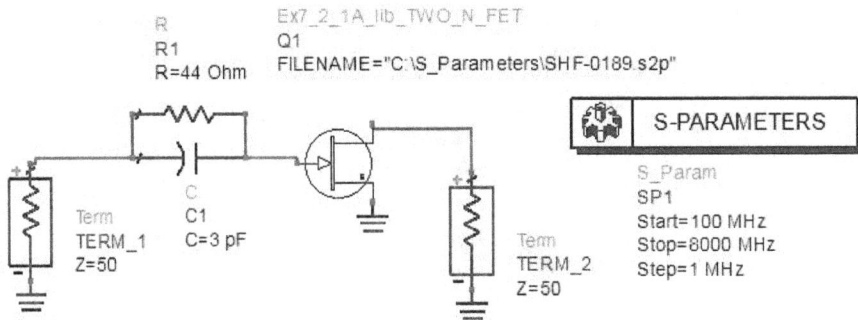

Figure 7-4 Added parallel RC stability network

Plot the parameters K, B1, and Gmax for the SHF-0189 device as shown in Figure 7-5.

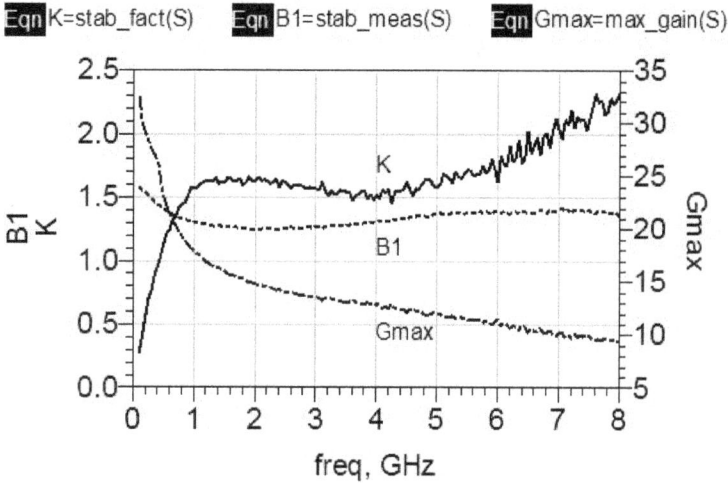

Figure 7-5 Stability parameters and GMAX with parallel RC network

As seen in Figure 7-5, the stability measure, B1, is greater than zero throughout the entire frequency range. However the stability factor, K, goes below one at frequencies from less than 427 MHz. We want to make sure that the device is stable at all frequencies so that there is no possibility of any value of Γ_{ML} or Γ_{MS} that could make the device unstable at these out-of-band frequencies. Therefore add an R-L network in shunt with the input of the transistor for additional low frequency stability. This R-L network can be absorbed into the bias feed for the gate voltage of the transistor. Choose a fixed value of 50 Ω for the resistor. Select a starting value of 100 nH for the inductor and make this value tunable. This R-L network will load the previous R-C network so its values will need to be changed. The inductor's value can be reduced which in turn reduces the low frequency gain while not significantly reducing the in band gain at 2.35 GHz. Figure 7-6 shows the final stabilization network for the device.

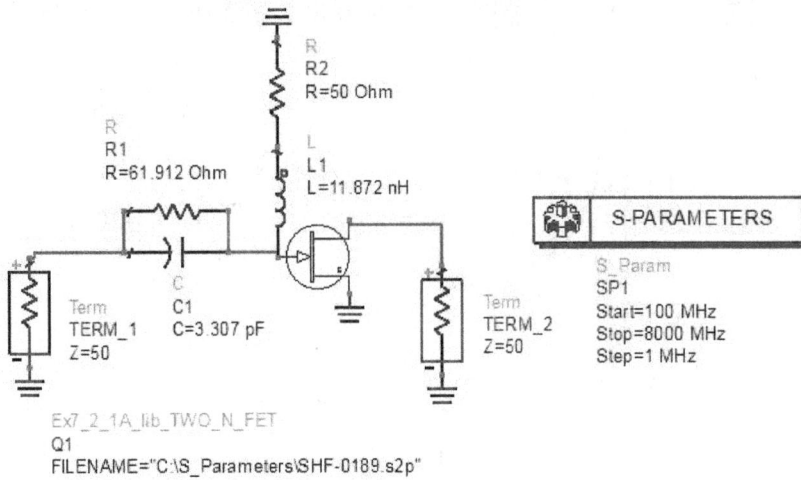

Figure 7-6 Stabilization network for the SHF-0189 transistor

Put a marker at 2350 MHz and notice in Figure 7-7 that, GMAX = 15.791 dB and K = 1.271 at the design frequency of 2.35 GHz. The device is now unconditionally stable for all frequencies from 50 MHz to 8 GHz.

Now that the device is unconditionally stable, we can determine the simultaneous conjugate match source and load impedance, $Z1_{sm}$ and $Z2_{sm}$.

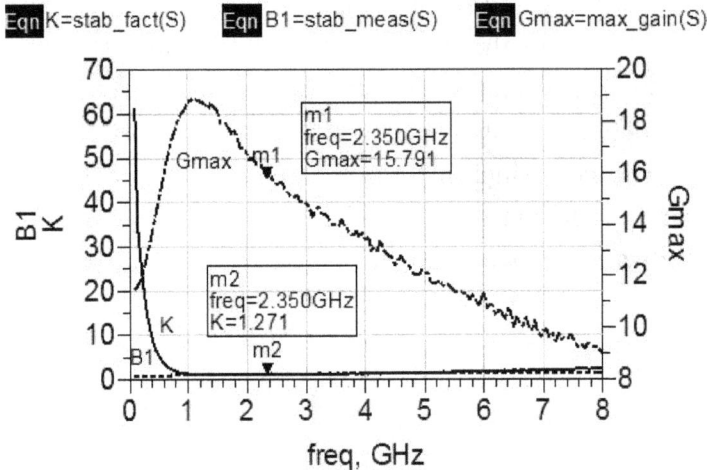

Figure 7-7 Stability parameters and G_{MAX} of the stabilized device

Simultaneous Match Reflection Coefficients and Impedances

ADS has built in functions for the calculation of the G1$_{sm}$ and G2$_{sm}$ simultaneous match reflection coefficients as well as the corresponding simultaneous match impedances Z1$_{sm}$ and Z2$_{sm}$. Using the schematic of Figure 7-6 create a tabular output of the simultaneous conjugate source and load parameters at 2350 MHz, as shown in Figure 7-8.

| Eqn Gmax=max_gain(S) | Eqn Z1sm=sm_z1(S) | Eqn Z2sm=sm_z2(S) | Eqn G1sm=sm_gamma1(S) | Eqn G2sm=sm_gamma2(S) |

freq	Gmax	Z1sm	Z2sm	G1sm	G2sm
2.350 GHz	15.791	6.918 + j22.023	13.720 + j16.471	0.793 / 131.771	0.605 / 141.090

Figure 7-8 Table showing G$_{MAX}$, Z1sm, Z2sm, G1sm, and G2sm

Note the equivalence between the ADS function notation and the respective reflection coefficient and impedance. The following numbers are used in the design of the impedance matching networks.

$$Z1sm = Z_{MS} = 6.92 + J22.02 \ \Omega \qquad\qquad G1sm = \Gamma_{MS} = 0.793 \angle 131.7$$

$$Z2sm = Z_{ML} = 13.72 + J16.47 \ \Omega \qquad\qquad G2sm = \Gamma_{ML} = 0.605 \angle 141.1$$

Input Matching Network Design

In the design of the SHF-0189 amplifier, we will use the analytical impedance matching techniques developed in chapter 5 and the L-Network Synthesis Utility to generate the matching networks.

When performing the analytical match we are matching the 50 Ω source impedance to the impedance looking into the device impedance (Z$_{IN}$ or Z$_{OUT}$ of Figure 7-7; not the Z$_{MS}$ or Z$_{ML}$. Z1 = Z$_{IN}$ and Z2 = Z$_{OUT}$ is the impedance looking into the input and output of the device. Z$_{MS}$ and Z$_{ML}$ are the source and load impedances that the device is looking into.

Therefore the complex impedances that we are matching are given as the conjugate of Z_{MS} and Z_{ML}.

$$Z1 = Z1^* sm = 6.92 - j22.02 \ \Omega$$
$$Z2 = Z2^* sm = 13.72 - j16.47 \ \Omega$$

To match the complex load impedance to a resistive source first calculate r and x by normalizing the load impedance with respect to the source resistor, and then determine the matching networks on the basis of the conditions summarized in Table 5-1. The conditions are repeated here.

- If $r < 1$ and $r^2 + x^2 - r < 0$ there exists only two matching networks obtained from Equations (5-21) through (5-24) in the Appendix B.

- If $r < 1$ and $r^2 + x^2 - r > 0$ there exists four matching networks obtainable from Equations (5-21) through (5-24) and Equations (5-26) through (5-29) in the Appendix B.

- If $r > 1$ there exists only two matching networks obtained from Equations (5-26) through (5-29) in the Appendix B.

Example 7.3-1: Design the maximum gain amplifier input matching network.

Solution: For the stabilized SHF-0189 input matching network, $r = 0.1384$ and $x = -0.44$, therefore, $r < 1$ and $r^2 + x^2 - r = 0.074 > 0$. According to the above conditions, there exists four L-networks that can be used for the input matching network. The following procedure shows how to use MATLAB script to calculate the element values of the second matching network.

- Enter the source and load impedance and design frequency
- Normalize the load impedance
- Write the appropriate equations to calculate B and X
- Based on the positive or negative values of B and X, calculate the element values of the matching network.

1. Enter design parameters and normalize the load impedance

Z0=50; RL=6.918; XL=-22.03; f=2.35e9, r=RL/Z0, x=XL/Z0

2. Calculate matching element values from Equations (5-34) and (5-35) in the Appendix B.

B2=-sqrt((1-r)/r)/Z0

X2=-Z0*(sqrt(r*(1-r))+x)

L1=X2/(2*pi*f)

L2=-1/(2*pi*f*B2)

The calculation results show that the series element is an inductor L1= 0.322 nH and the shunt element is another inductor L2=1.357 nH.

To display the response of the matching network, create a new workspace in ADS and open a new schematic window. Insert the S_Params Template and connect the two inductors as shown in Figure 7-9.

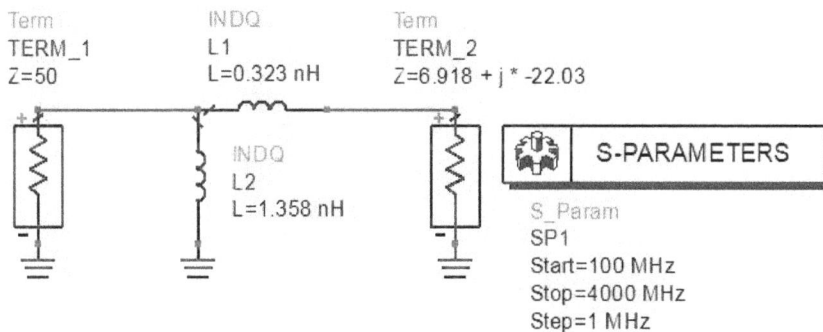

Figure 7-9 Impedance matching network for the input match

Simulate from 1350 MHz to 3350 MHz and display S11 and S21, in dB, in a rectangular plot, as shown in Figure 7-10.

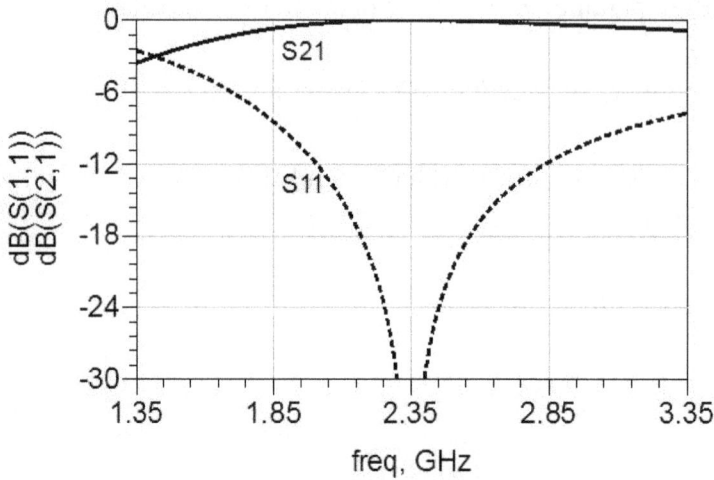

Figure 7-10 Response of the Impedance matching network

Output Matching Network Design

Example 7.3-2: Design the output matching network for the SHF-0189 amplifier at 2350 MHz.

Solution: For the stabilized SHF-0189 output matching network we use the second matching network, as shown in Figure 7-11.

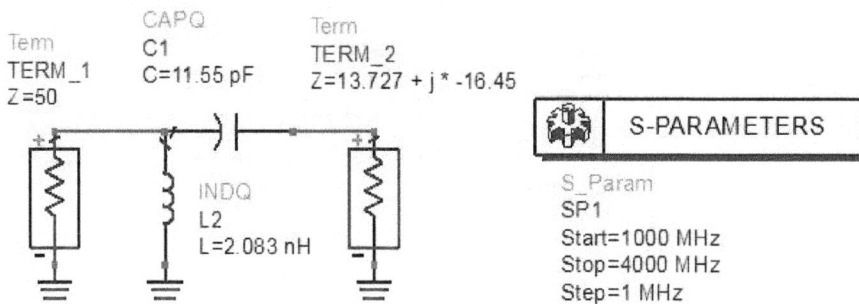

Figure 7-11 Output matching L-network

The simulated response is show in 7-12

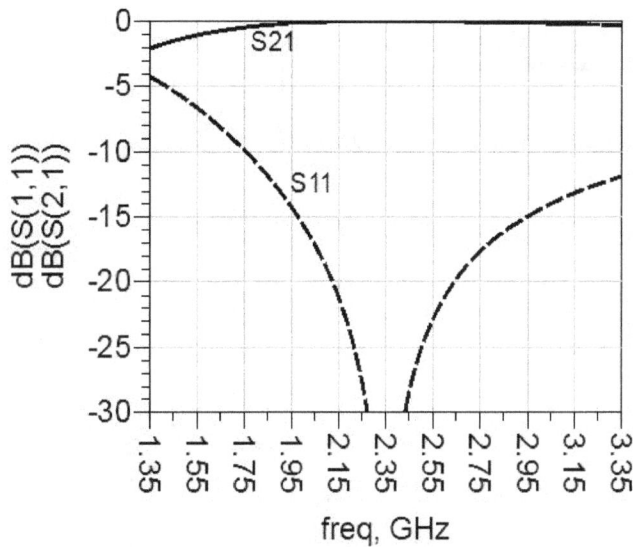

Figure 7-12 Response of the output matching L-network

Ideal Model of the Maximum Gain Amplifier

Example 7.3-3: Assemble, simulate, and display the amplifier response.

Solution: Create the ideal model of the amplifier by attaching the input and output matching networks to the device's S parameter file based model as shown in Figure 7-13. Use a Linear Sweep from 100 MHz to 5000 MHz to analyze the response of the amplifier. The swept gain, input return loss, and output return loss is shown in Figure 7-14. As the return loss shows, an excellent match has been achieved at the input and output ports. Also the maximum gain (15.791 dB) of the amplifier at 2.35 GHz is exactly the originally calculated G_{MAX} indicating a successful simultaneous conjugate impedance match.

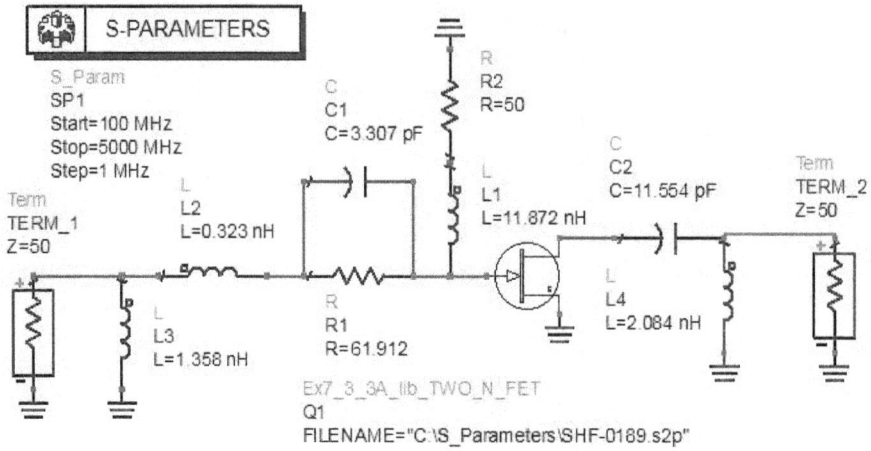

Figure 7-13 Schematic of the ideal amplifier circuit

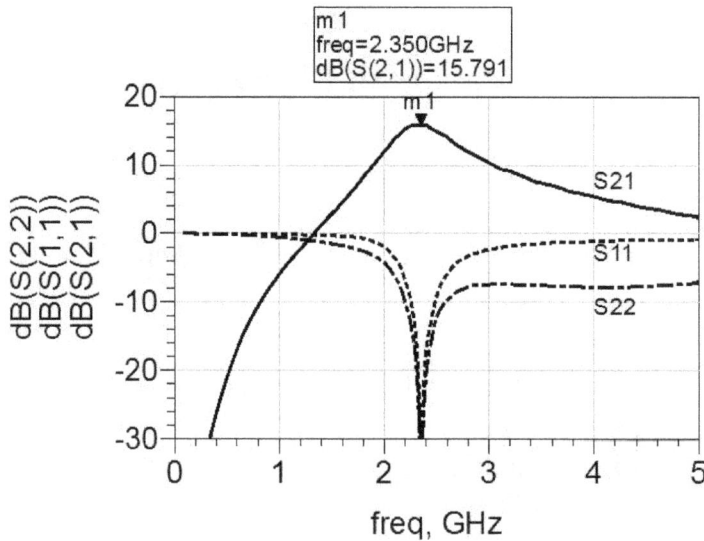

Figure 7-14 Swept response of the ideal amplifier circuit

The fractional bandwidth of the amplifier is calculated by sweeping the response over a narrower frequency range as shown in Figure 7-15.

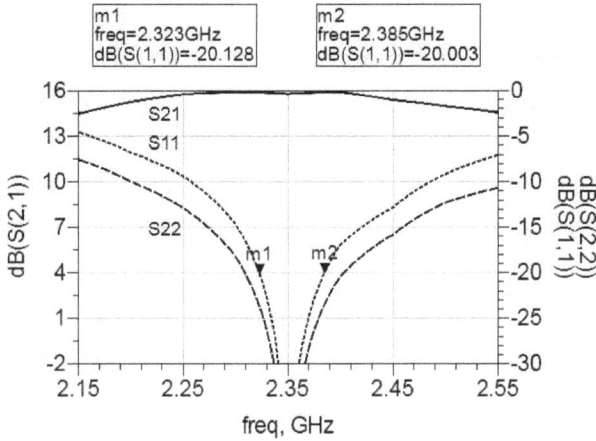

Figure 7-15 Ideal amplifier response with markers at 20 dB return loss

Markers are placed on the passband response at the 20 dB return loss corresponding to 0.1 dB ripple bandwidth.

$$BW_{20dB} = 2385\text{-}2323 = 62 \text{ MHz}.$$

This corresponds to a fractional bandwidth of about 2.66 % consistent with a narrowband amplifier that is matched at the single frequency of 2.35 GHz.

$$FBW_{20dB} = \frac{62}{\sqrt{(2385).(2323)}} = 2.66\%$$

Amplifier Physical Design and Layout

Example 7.4-1: Design the physical amplifier bias feed and the physical PCB layout for the SHF-0189 maximum gain amplifier.

Solution: Start the physical design of the amplifier by placing the device S parameter file on the schematic. Use a microstrip via hole model for the connection of the transistor source lead to ground. Specify a via hole radius of 12 mils. Then assign the artwork replacement element for the SHF-0189 by opening the Properties window of the S parameter file. Select the Change Footprint button and browse to the location of the artwork replacement element created in the previous section. Proceed with the construction of the stabilization network portion of the design. Use Modelithic surface mount components (30 mil x 60 mil) for the stabilization components as these will be easy to tune or optimize Use the S parameter files of the components that we do not intend to optimize. A 470 pF capacitor is added as a gate bypass capacitance. On the drain side of the transistor a 33 nH inductor is used as an RF choke to supply the drain voltage. Edit the footprint of the S parameter files and change the footprint to a 0603 surface mount component from the ADS library. The Modelithic components will automatically have the correct footprint assigned to its properties.

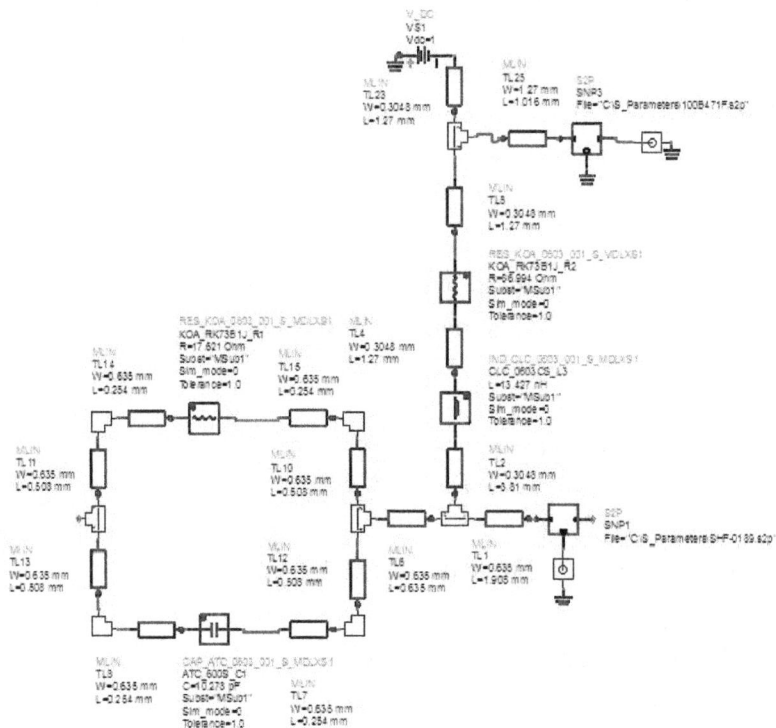

Figure 7-16 Construction of physical amplifier bias feed

Continue with the physical model construction by adding the matching networks. Note that on the input side of the circuit the matching network has an inductor connected to ground. Because there will be a negative voltage applied to the gate of the transistor, a DC blocking capacitor must be added to keep the inductor from shorting out the gate bias. Add an S parameter file for a 1.5 pF chip capacitor to act as a DC block. The capacitance value should be chosen such that the capacitor is operating in series resonance for minimum insertion loss. Knowing that the chip capacitor typically has some series inductance, the small series inductance of 0.323 nH can be absorbed into the DC blocking capacitor. The output matching circuit includes a series capacitance so no additional DC blocking capacitor in needed. The completed schematic of the physical amplifier model is shown in Figure 7-17. The physical PCB layout is shown in Figure 7-18.

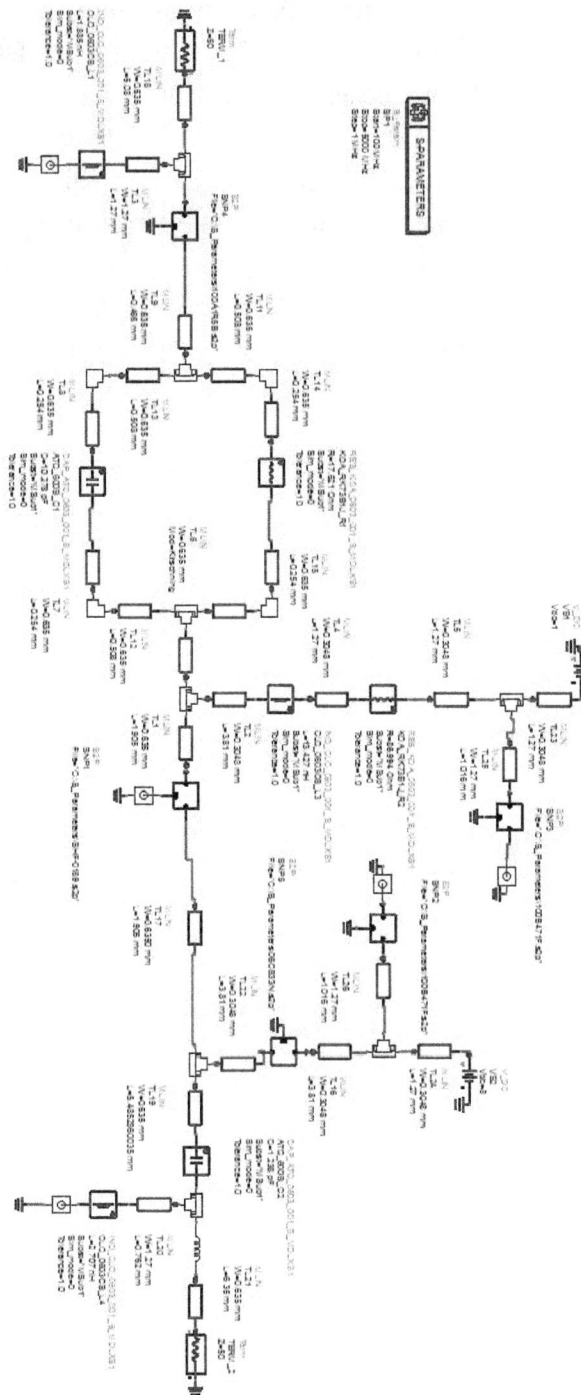

Figure 7-17 Schematic of the physical amplifier model

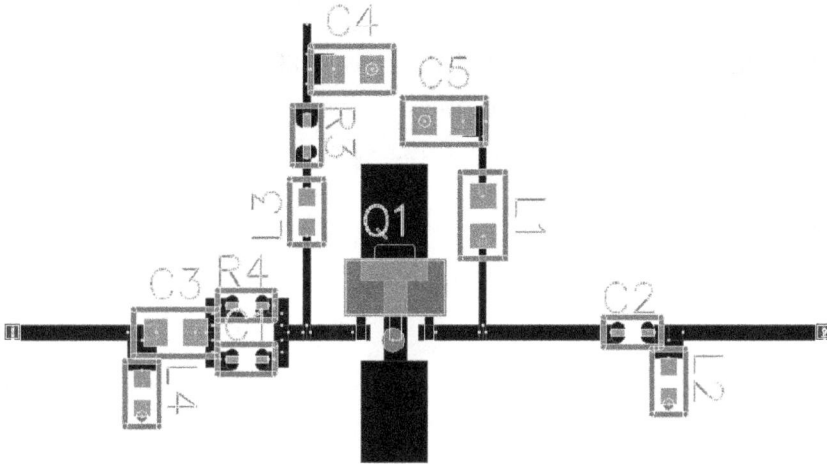

Figure 7-18 Physical layout of completed amplifier circuit

Optimization of the Amplifier Response

After completion of the physical circuit schematic of Figure 7-18 the circuit response is swept from 100 MHz to 5000 MHz. The response of the amplifier significantly changes from the response of the ideal amplifier response of Figure 7-14. The gain and return loss have shifted in frequency. This is a typical result of replacing ideal elements with real physical models. This ability to deal with package parasitics and physical layout is a very powerful benefit of modeling the amplifier in ADS. We know that it is possible to make the component element values tunable and manually tune each element while observing the change in response. This is quite useful and encouraged so that you can see the circuit sensitivity to particular components. As the number of components and tunable elements increase in a given design, the tuning of individual elements can be cumbersome. In this section we will introduce the use of Optimization to tune the amplifier's response to the desired characteristics. We can think of Optimization as an automatic circuit tuner. Selected components in the amplifier can be assigned as tunable variables. Multiple goals can then be set such as gain levels, return loss, etc. The optimization is then run and the ADS software will execute a mathematical optimization algorithm to determine the values of the variables required to achieve the specified goals. In this example set all seven of the Modelithic model values as tunable. Also make the microstrip line lengths for the series sections TL9 and TL19 variable. We will let the software optimize these variables to achieve the desired response. Make sure to set response goals for out-of-band responses as well as in band response.

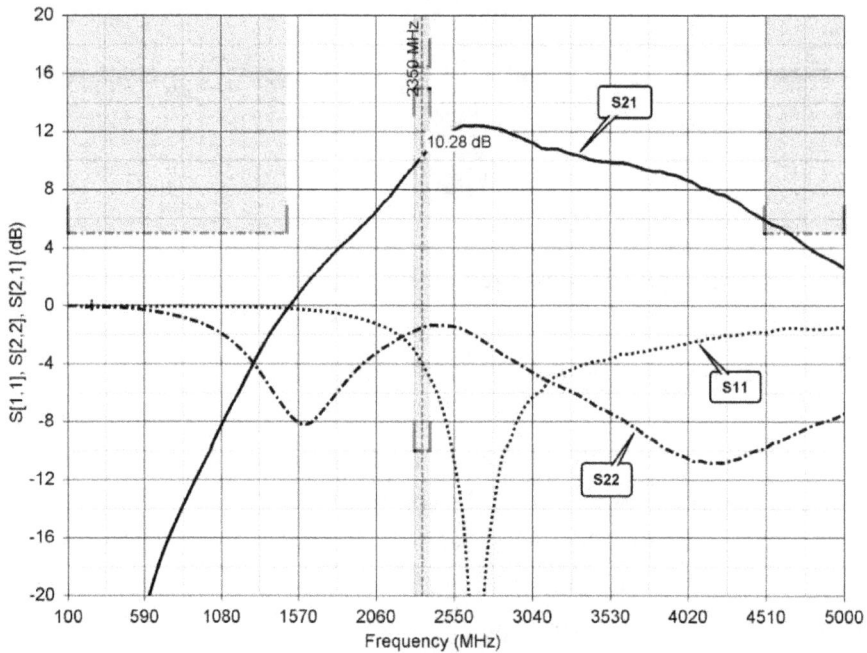

Figure 7-19 Initial response of the physical amplifier circuit of Figure 7-17

Specific Gain Amplifier Design

In the design of RF and microwave amplifiers, it is common to design with potentially unstable transistors. The designer must know how to deal with potentially unstable devices. This section outlines the design of a 2.3 GHz amplifier using the RT243 GaN HEMT device. The RT243 has a usable frequency range of 100 MHz to 5 GHz.

Specific Gain Design Example

Example 7.5-1: Analyze the stability condition for the RT243 device.

Solution: Examine the stability circles for the RT243 by creating a new workspace and a new schematic in ADS with the small signal S parameter file as shown in Figure 7-20. Setup an S parameter simulation with both the start and stop frequencies set at 2.3 GHz.

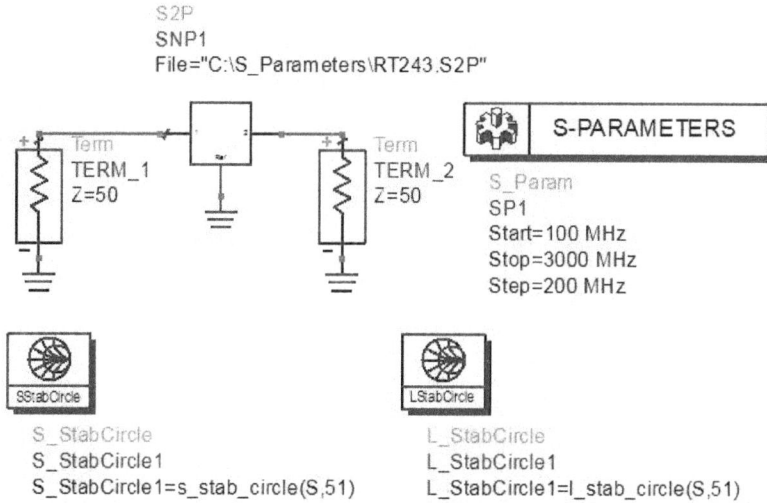

S2P
SNP1
File="C:\S_Parameters\RT243.S2P"

Term
TERM_1
Z=50

Term
TERM_2
Z=50

S-PARAMETERS

S_Param
SP1
Start=100 MHz
Stop=3000 MHz
Step=200 MHz

SStabCircle

S_StabCircle
S_StabCircle1
S_StabCircle1=s_stab_circle(S,51)

LStabCircle

L_StabCircle
L_StabCircle1
L_StabCircle1=l_stab_circle(S,51)

Figure 7-20 Schematic with RT243 S parameter file

A sweep of the circuit will result in a plot of the center of each stability circle displayed. To display the entire circle, click on the center trace (dot) on the Smith Chart. As Figure 7-21 shows the circles are far away from the center of the chart therefore making the outside of the circles the stable region. There is only a small region of the output stability circle that represents an unstable region of impedance. These unstable regions are shaded as shown in Figure 7-21. If the device were unconditionally stable the stability circles would lie completely outside the circumference of the Smith Chart. Add another simulation from 100 MHz to 3000 MHz with 100 MHz steps. Then add a Table to display K, B1, and GMAX. Note that stability factor k is less than 1 indicating that the transistor is potentially unstable over its entire frequency range. G_{max} for the device without a stabilization network is 22.6 dB at 2.3 GHz. We want to design the amplifier to have a gain of 16 dB with a 50 Ω source and load impedance.

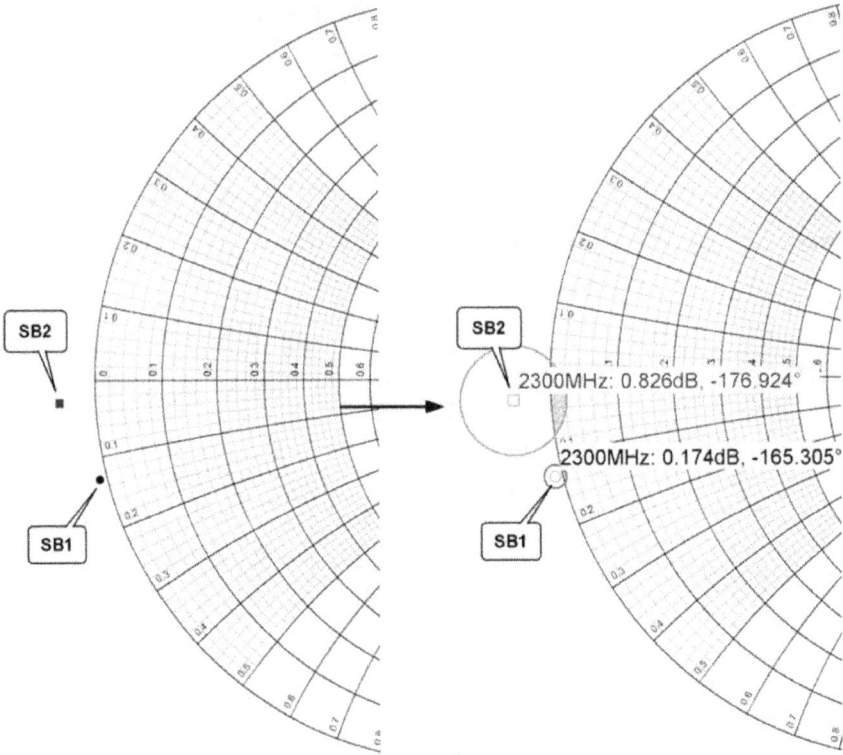

Figure 7-21 Stability circle centers and full circles at 2.3 GHz

The stability parameters are shown in Table 7-2.

Eqn B1=stab_meas(S)   Eqn K=stab_fact(S)   Eqn Gmax=max_gain(S)

freq	B1	K	Gmax
100.0 MHz	0.966	0.272	35.753
300.0 MHz	0.838	0.122	30.862
500.0 MHz	0.904	-0.236	28.363
700.0 MHz	0.853	0.266	27.462
900.0 MHz	0.845	0.316	26.400
1.100 GHz	0.826	0.352	25.528
1.300 GHz	0.800	0.379	24.778
1.500 GHz	0.769	0.419	24.125
1.700 GHz	0.739	0.430	23.546
1.900 GHz	0.711	0.463	23.035
2.100 GHz	0.690	0.537	22.750
2.300 GHz	0.666	0.654	22.610
2.500 GHz	0.641	0.802	22.448
2.700 GHz	0.625	0.861	22.188
2.900 GHz	0.609	0.931	21.911
3.000 GHz	0.601	0.970	21.765

Table 7-2 RT243 device stability factors and GMAX

Fortunately ADS has the ability to plot a series of constant gain circles. This enables the plot of the circles on the Smith Chart without the need to solve any equations. Selecting the GP function (power gain circles) in the output graph plots a series of seven gain circles. When K < 1 the smallest circle represents the maximum gain, GMAX, and the inside of the circle is shaded. This means that impedances inside this circle can be unstable. The successive Power Gain circles represent gain values of GMAX -1dB, GMAX-2dB, GMAX-3 dB, GMAX-4 dB, GMAX-5 dB, and GMAX-6 dB. Thus the circles represent the gain values of: 22.6 dB, 21.6 dB, 20.6 dB, 19.6 dB, 18.6 dB, 17.6 dB and 16.6 dB respectively. Figure 7-22 shows that the last or largest circle is the 16.6 dB power gain circle. An examination of the power gain and stability circles reveals another reason why we may not want to use the maximum gain from the device. The highest gain circle is very close to the region of instability on the output stability circle, SB2.

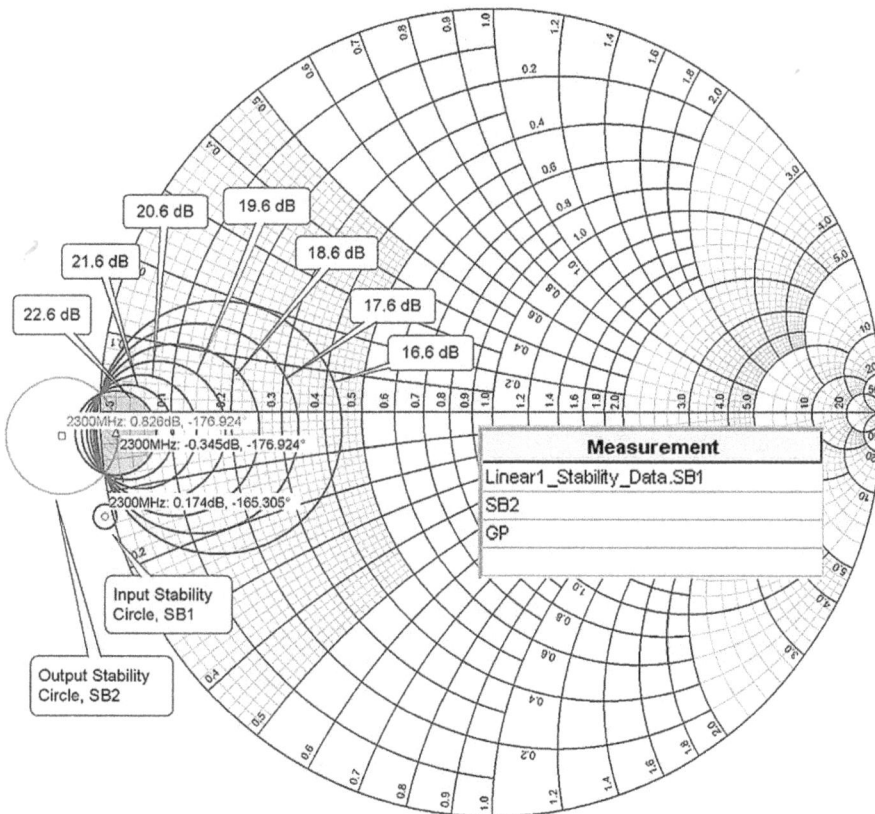

Figure 7-22 Constant power gain circles for the RT243 at 2.3 GHz

We'll choose the 16.6 dB circle because it is the closest to the design goal of 16 dB. The load impedance can be any point on the circumference of the 16.6 dB circle that does not enter the shaded, unstable, region. It is a good practice to choose locations that are not close to the unstable regions of the stability circles. The load impedance can be visually read from the Smith Chart. The load impedance selected on the 16.6 dB Power Gain Circle and the real axis is $22 + j0\ \Omega$.

Once the load impedance has been selected it is necessary to calculate the resulting source reflection coefficient that will achieve a conjugate match on the input. The following equation is used to calculate the source reflection coefficient.

$$\Gamma_S = \left[S_{11} + \frac{S_{12}S_{21}\Gamma_L}{1-(\Gamma_L S_{22})} \right]^*$$

The chosen load impedance must first be converted to a reflection coefficient, Γ_L. This is easily handled by the following MATLAB equations. The source impedance is calculated as 0.943-j6.594 Ohms. Solution of Γ_S is given here.

1. Enter design parameters

RL=22, XL=0

Load=Linear2_Data.S[1,1]

Num=Linear1_Data.S[2,1].*Linear1_Data.S[1,2].*Load

Denom=1-(Load.*Linear1_Data.S[2,2])

Source=Linear1_Data.S[1,1].+(Num/Denom)

2. Calculate the conjugate of the Source Impedance

Zsource=gammatoz(50,Source)   %polar to rectangular

Zsource_real=real(Zsource[1])

Zsource_imag=-imag(Zsource[1])   %make conjugate

Next enter the values of Zsource_real and Zsource_imag to perform the Input Matching Circuit Design

Design of the Impedance Matching Networks

The simple two-element LC matching circuit can be synthesized for both the input and output matching circuits.

Example 7.5-2: Design the output matching network for the RT243 transistor at 2300 MHz.

1. Enter design parameters and normalize the load impedance

Z0=50; RL=22; f=2.3e9, r=RL/Z0

2. Use the equations in the Appendix B to calculate the element values

B1=sqrt((1-r)/r)/Z0

X1=Z0*sqrt(r*(1-r))

L1=X1/(2*pi*f)

C2=B1/(2*pi*f)

The calculation results show that L1 = 1.717 nH and C2 = 1.561 pF, as shown in Figure 7-23.

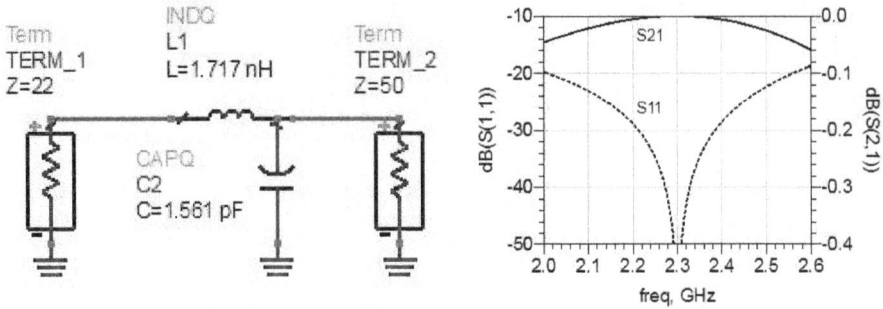

Figure 7-23 Schematic used to design the output matching circuit

Example 7.5-3: Design the input matching network for RT243 transistor at 2300 MHz.

To design the input matching network, we use the input impedance of the RT243 transistor.

1. Enter design parameters and normalize the load impedance

Z0=50; RL=0.943; XL=6.594; f=2.3e9, r=RL/Z0, x=XL/Z0

2. Use the equations in Appendix B to calculate the element values

B1=sqrt((1-r)/r)/Z0

X1=Z0*sqrt(r*(1-r))-x*Z0

L1=X1/(2*pi*f)

C2=B1/(2*pi*f)

The calculation results show that L1 = 0.014 nH and C1 = 9.982 pF, as shown in Figure 7-24.

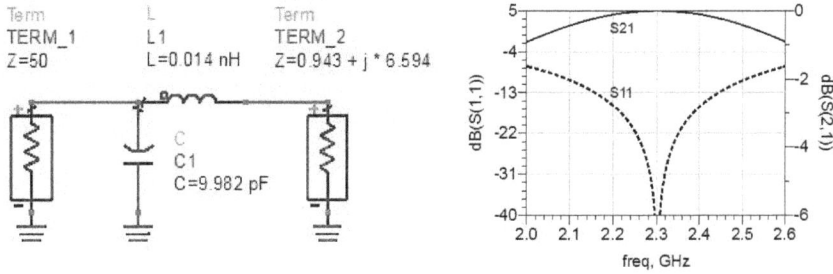

Figure 7-24 Schematic used to design the input matching circuit

Assembly and Simulation of the Specific Gain Amplifier

Example 7.5-4: Assemble, simulate, and display the response of the specific gain Amplifier.

Solution: Create a new schematic and attach the input and output matching circuits to the device. Create a linear analysis in ADS and sweep the amplifier from 1.3 GHz to 3.3 GHz. The resulting response is shown in Figure 7-25.

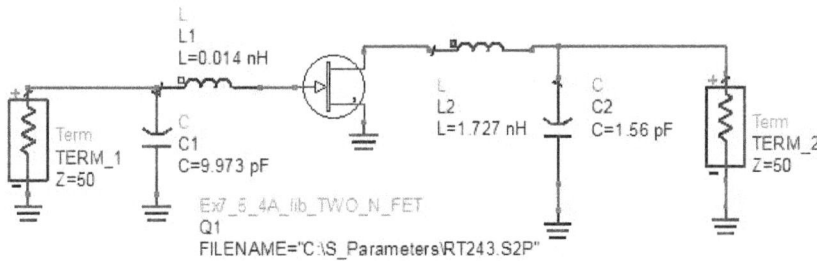

Figure 7-25 Ideal design of the specific gain amplifier

The simulated response of the specific gain amplifier is shown in Figure 7-26.

m1
freq=2.300GHz
dB(S(2,1))=16.621

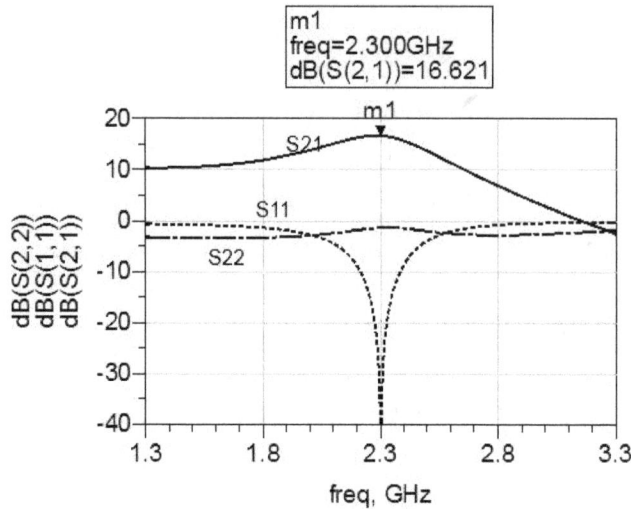

Figure 7-26 Response of the power gain matched amplifier

Note that the gain at 2300 MHz is 16.64 dB verifying that the Power Gain circle match has been successful. Also note a major difference in return loss compared to the simultaneous conjugate match case. Here we see that the output return loss is rather poor, -1.39 dB. This is a result of the selective mismatching process to achieve a much lower value of gain than the GMAX of the device. The input return loss is very good because a conjugate match of the resulting device and load impedance combination was performed.

Low Noise Amplifier Design

One important case of selectively mismatched amplifier design is the Low Noise Amplifier, LNA. In LNA design the input is not matched to the reflection coefficient that results in maximum gain but rather to a reflection coefficient that gives the desired noise figure. Low noise amplifiers are frequently used as the input stage in a radio receiver or satellite down converter to minimize the noise that is added to the amplified signal.

Low Noise Amplifier Design

This section outlines the basic steps that are required to design a single stage Low Noise Amplifier for a UHF satellite downlink. The amplifier is intended to operate with a source and load impedance of 50 Ω. The design specifications are given as:

Center Frequency:	432 MHz
Gain:	20 dB minimum
Noise Figure:	1.2 dB maximum
Output Return Loss:	Less than -10 dB

The Avago AT30511 low noise transistor will be used for this design. The S parameter and noise parameter file for this device is shown in Figure 7-27.

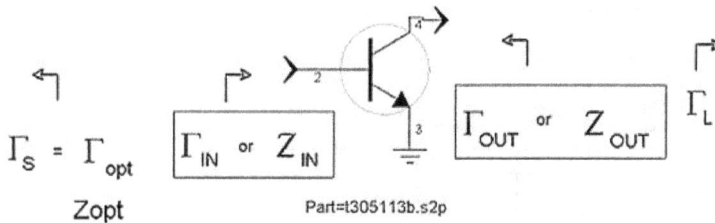

Figure 7-27 AT30511 with input and output matching

Example 7.6-1: Calculate the stability and noise parameters for the AT 30511 device.

Solution: Create a schematic in ADS with the device S parameter file as shown in Figure 7-28. Make sure to select **yes** for calculating noise and setting **Temp** to IEEE standard of 16.85 degrees.

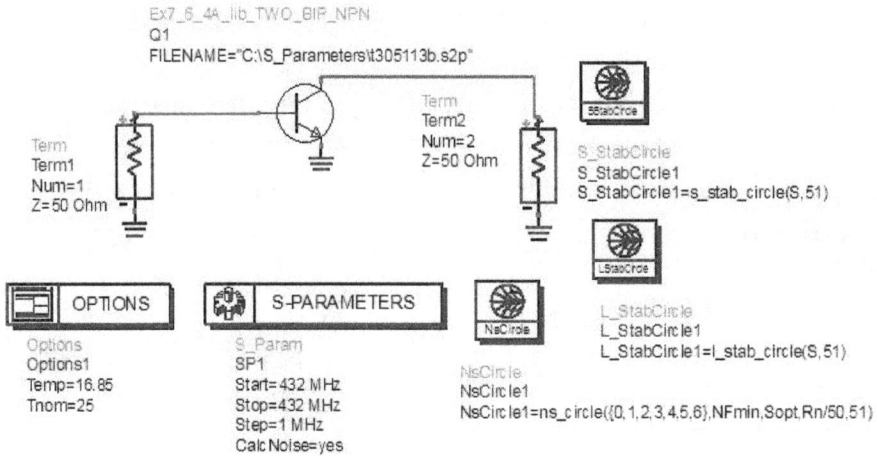

Ex7_6_4A_lib_TWO_BIP_NPN
Q1
FILENAME="C:\S_Parameters\t305113b.s2p"

Term
Term2
Num=2
Z=50 Ohm

S_StabCircle
S_StabCircle1
S_StabCircle1=s_stab_circle(S,51)

L_StabCircle
L_StabCircle1
L_StabCircle1=l_stab_circle(S,51)

Term
Term1
Num=1
Z=50 Ohm

OPTIONS

Options
Options1
Temp=16.85
Tnom=25

S-PARAMETERS

S_Param
SP1
Start=432 MHz
Stop=432 MHz
Step=1 MHz
Calc Noise=yes

NsCircle
NsCircle1
NsCircle1=ns_circle({0,1,2,3,4,5,6},NFmin,Sopt,Rn/50,51)

Figure 7-28 AT30511 with input and output matching

Add a Table to display the stability parameters K and B1 and the noise parameters Γ_{opt} and Z_{OPT}.

Eqn K = stab_fact(S)   Eqn B1 = stab_meas(S)   Eqn Gain = max_gain(S)

freq	K	B1	Gain	Sopt	NFmin
432.0 MHz	0.387	0.274	24.974	0.780 / 7.470	1.083

Figure 7-29 Parameters K, B1, Γ_{opt}, NFmin and Gmax at 432 MHz

Figure 7-29 shows that K < 1 meaning that the device is potentially unstable. This means that there are source and load reflection coefficients that can result in instability and oscillation.

Also add a Smith Chart with S_{opt} and the stability circles. Then create a Smith Chart and display the stability circles along with the noise circles.

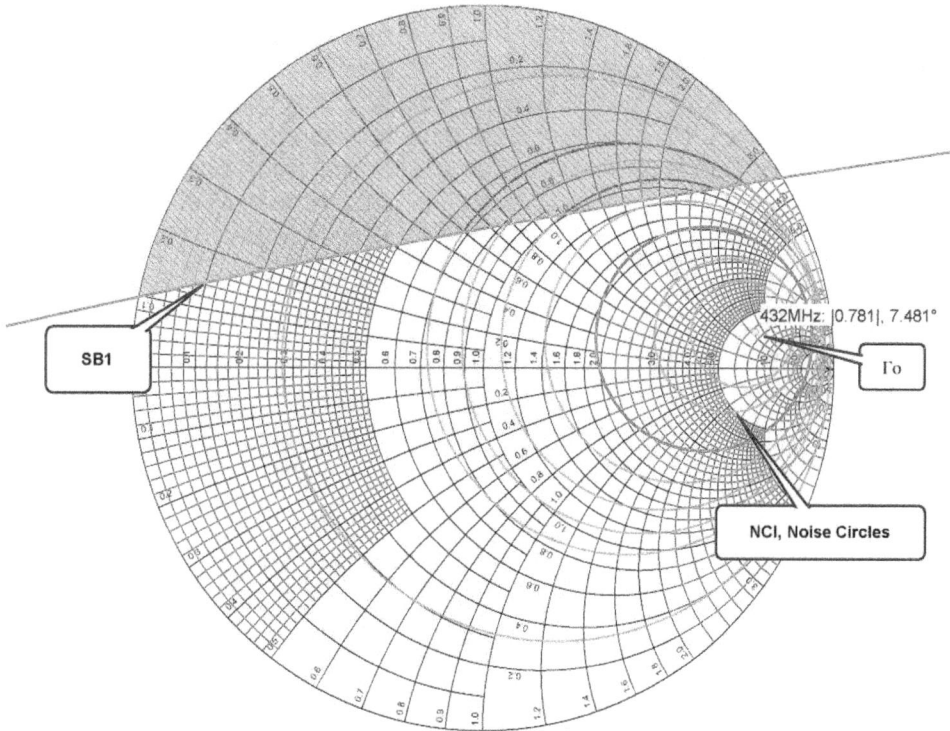

Figure 7-30 Input stability circle with noise circles

Figure 7-30 shows a plot of the input stability circle along with the noise circles and S_{opt} . In this example, the inside of the input stability circle, SB1, represents the region of stable source reflection coefficients. ADS shades the region of the stability circle that represents unstable reflection coefficients. Because the input reflection coefficient, S_{opt} , is well inside of the input stability circle, SB1, and we have no specification for input return loss, we can proceed with matching the input to S_{opt}. From the discussion of Specific Gain matching we can deduce that because a specific reflection coefficient is selected at the input of the device, the available gain circles can be used to determine the expected gain from the device when matched to S_{opt}.

Figure 7-31 shows the available gain circles for the AT30511 at 432 MHz plotted along with S_{opt}. The available gain circles are drawn for gains of 0, 1, 2, 3, 4, 5 and 6 dB less than maximum gain. The 21.9 dB circle intersects

S_{opt} therefore the expected gain of the device when matched to S_{opt} is 21.9 dB.

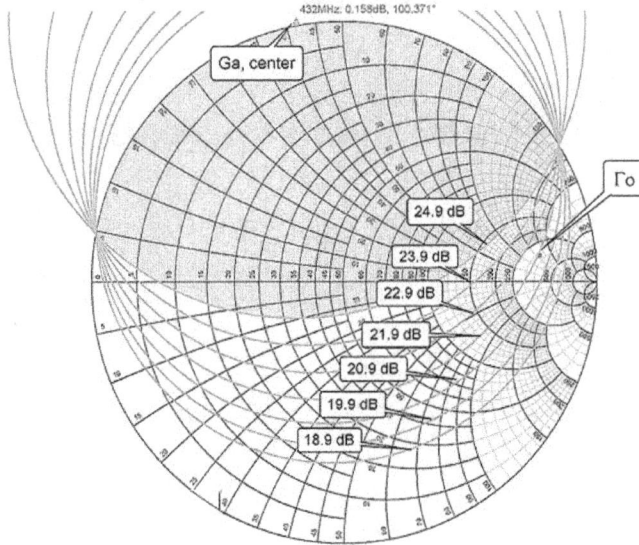

Figure 7-31: Available gain circles and Γ_{opt}

LNA Input Matching Network Design

The input and output matching circuits will be analytically designed using the single-stub matching networks. Because we are designing essentially a fixed frequency LNA, a simple single-stub matching network can be used.

Example 7.6-2 Design a line and stub input matching network for the low noise amplifier using AT30511 device at 432 MHz.

Solution: We match the 50 Ω source impedance to the input impedance Z_{in} of the device. Z_{opt} is the impedance looking from the device back toward the 50 Ω source. Therefore, in the design of the input matching network, we must use the conjugate of Z_{opt} for the complex load impedance. The calculation gives $Z_{opt} = 318.4 + j166.1$,Ω therefore, Z_{in} is defined as $Z_{opt}{}^* = 318.4 - j166.1$ Ω.

1. Enter design parameters and normalize the load

RS=50; RL=318.6; XL=-166.1; f=432e6, r=RL/RS, x=XL/RS

2. Use the equations in Appendix C to calculate t1, t2, d1, d2, B1, B2, so1 and so2

t1=(x+sqrt(r*(r^2+x^2-2*r+1)))/(r-1)

t2=(x-sqrt(r*(r^2+x^2-2*r+1)))/(r-1)

d1=360*(pi+atan(t1))/(2*pi)

d2=360*(pi+atan(t2))/(2*pi)

B1=(x*t1^2+(r^2+x^2-1)*t1-x)/(RS*(r^2+x^2+t1^2+2*x*t1))

B2=(x*t2^2+(r^2+x^2-1)*t2-x)/(RS*(r^2+x^2+t2^2+2*x*t2))

so1=360*(pi-atan(RS*B1))/(2*pi)

so2=-360*atan(RS*B2)/(2*pi)

Calculation results show that d1 = 66.945, d2 = 105.58, so1 = 111.784, and so2 = 68.216. For the input matching network,

As the calculations show there are two single-stub matching networks that match the 50 Ωsource impedance to the load impedance. We select the shorter open-stub length so2 = 68.216 degrees and the corresponding line d2 = 105.58 degrees, as shown in Figure 7-32.

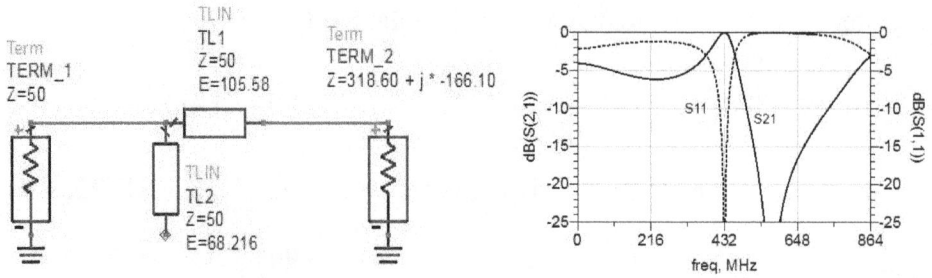

Figure 7-32 Single-stub input matching network and simulated response

LNA Output Matching Network Design

The output load reflection coefficient, Γ_L, is defined in terms of the selected source reflection coefficient, Γ_S, as given by the equation.

$$\Gamma_L = \left(S_{22} + \frac{S_{12}S_{21}\Gamma_S}{1 - S_{11}\Gamma_S} \right)^*$$

This equation can be solved by letting $\Gamma_S = \Gamma_{opt}$ and calculating the load reflection coefficient, Γ_L.

To use the impedance matching utility we need the conjugate of Z_L or Z_{OUT} of the device or in this case: $Z_{OUT} = 71.9 - j54.4\ \Omega$. Check the location of Γ_L with respect to the output stability circle to make sure that the impedance does not lie in the region of instability. Figure 7-33 shows the location of Γ_L with respect to the output stability circle, SB2.

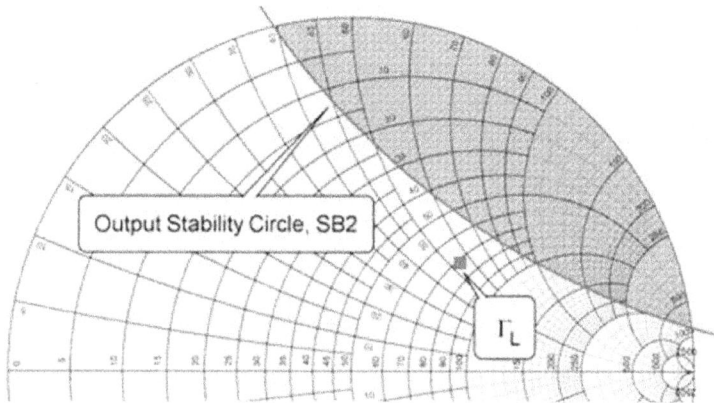

Figure 7-33 Location of Γ_L and the output stability circle

The impedance corresponding to Γ_L can be entered into an impedance element and plotted on the same graph as the output stability circle. The inside of the output stability circle is shaded which means that this area represents unstable output load reflection coefficients. Because Γ_L is safely outside of the SB2 circle we can use Γ_L as an acceptable output reflection coefficient.

Example 7.6-3 Design a single-stub output matching network for the low noise amplifier using AT30511 device at 432 MHz.

Solution: Now we can design a single-stub output impedance matching network. The procedure follows.

1. Enter design parameters and normalized load impedance

RS=50; RL=71.9; XL=-54.4; f=432e6, r=RL/RS, x=XL/RS

2. Use the equations in Appendix C to calculate t1, t2, d1, d2, B1, B2, so1, and so2

t1 = (x+sqrt(r*(r^2+x^2-2*r+1)))/(r-1)

t2 = (x-sqrt(r*(r^2+x^2-2*r+1)))/(r-1)

$$d1 = 360*(atan(t1))/(2*pi)$$

$$d2 = 360*(pi+atan(t2))/(2*pi)$$

$$B1 = (x*t1^2+(r^2+x^2-1)*t1-x)/(RS*(r^2+x^2+t1^2+2*x*t1))$$

$$B2 = (x*t2^2+(r^2+x^2-1)*t2-x)/(RS*(r^2+x^2+t2^2+2*x*t2))$$

$$so1 = 360*(pi-atan(RS*B1))/(2*pi)$$

$$so2 = -360*atan(RS*B2)/(2*pi)$$

The calculation results show that there are two single-stub matching networks that match the 50 Ω load impedance to the LNA output impedance. For the output matching network, we select the shorter open-stub length so2 = 44.364 degrees and the corresponding line d2 = 99.959 degrees.

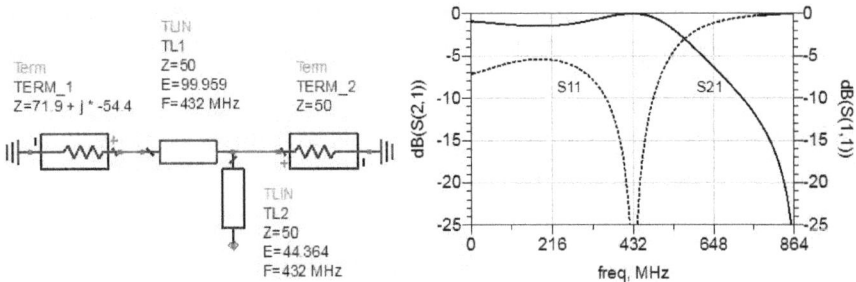

Figure 7-34 Output matching network and simulated response

Linear Simulation of the Low Noise Amplifier

Example 7.6-4: Assemble, simulate, and display the response of the low noise Amplifier.

Solution: Create a new schematic in ADS with the input and output matching networks attached to the device. Add a new S parameter simulation that sweeps the amplifier from 400 MHz to 500 MHz. Create a graph and plot the amplifier gain S21, as shown in Figure 7-35.

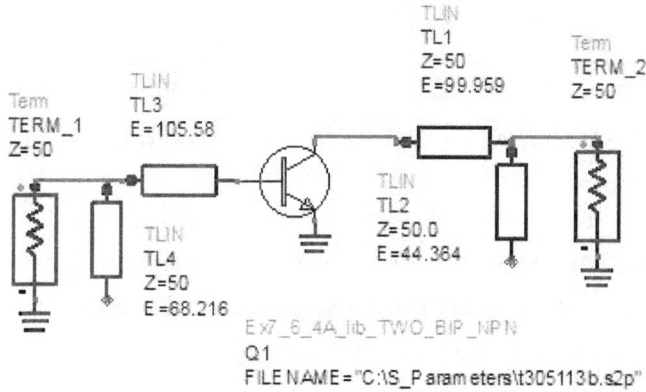

Figure 7-35 LNA schematic using single-stub matching networks

Simulate the schematic from 402 to 462 MHz and display S11, S22, and the gain, S21, in a rectangular plot.

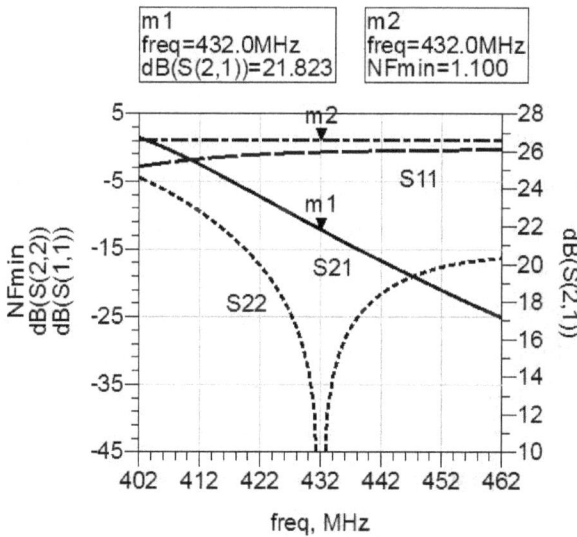

Figure 7-36 Marker showing LNA gain at 432 MHz

The reading at marker m1 shows that the amplifier gain at 432 MHz is 21.823 dB which is well within the acceptable range of the expected 21.9 dB. The simulated noise figure is 1.1 dB which is very close to the NFmin that was predicted in Figure 7-29.

Note that the Figure 7-35 is the ideal schematic of the LNA. The remaining design sequence should include the microstrip interconnecting lines and bias feeds. The circuit should then be optimized to bring the final response as close as possible to the ideal circuit results as was done for maximum gain amplifier.

Figure 7-36 also shows that the LNA input is selectively mismatched to achieve the minimum noise figure while the output is perfectively matched to achieve maximum gain.

Power Amplifier Design

The amplifier circuits described thus far have been designed using the measured S parameters that represent the active device. The S parameters are based on the small signal characteristics of the device. That is to say that the device is operating well below its maximum output power capability. This would suggest that the S parameters are defined for class A amplification, operating well within the linear portion of their power transfer characteristic. The S parameters can therefore be used to design an amplifier for specific values of gain but not output power. As we drive the transistor closer to its maximum output power the signal excursion is occurring over a much wider range of the transistor's load line. This is to say that the device is now operating under large signal conditions. There exists a specific source and load impedance into which the transistor can produce its maximum output power. Because the device S parameters are no longer defined under large signal conditions, they cannot be used to determine the optimum load impedance for maximum output power. These impedances are normally determined by performing a load pull measurement on the device.

Data Sheet Large Signal Impedance

In many cases high power devices are intended for specific applications such as WiMAX, cellular base station, or mobile radio. In these applications the device manufacturer may perform the load pull analysis at specific frequencies and present the source and load impedance as an equivalent series circuit on the data sheet. This allows the designer to treat the power matching process as a simple impedance matching exercise. Figure 7-37 shows an excerpt of a data sheet for the Nitronex NPT25100 GaN HEMT device. As the data sheet shows the optimum source and load impedances are given in tabular and Smith Chart forms.

Frequency (MHz)	Z_S (Ω)	Z_L (Ω)
2140	12.1 - j20.0	2.6 - j2.6
2300	10.0 - j3.0	2.5 - j2.3
2400	9.5 - j3.0	2.5 - j2.5
2500	9.0 - j3.0	2.5 - j2.7
2600	8.5 - j3.0	2.5 - j3.1
2700	8.0 - j3.0	2.5 - j3.3

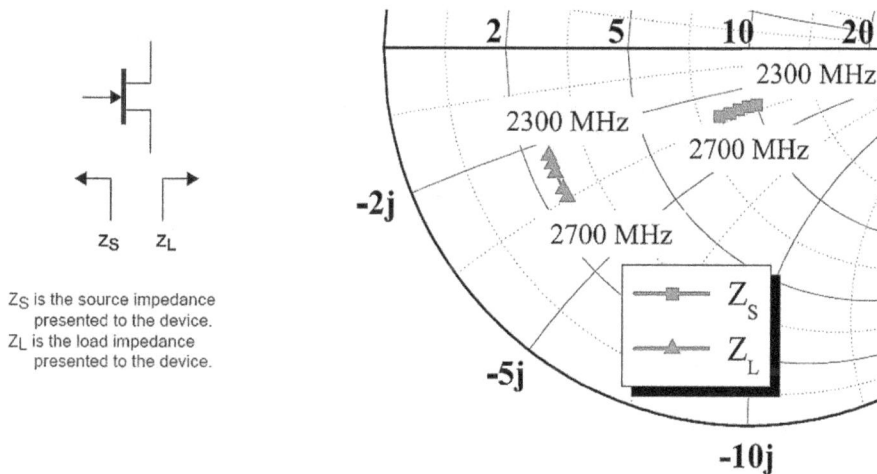

Figure 7-37 Optimum source and load impedance (*courtesy of Nitronex*)

As Figure 7-37 shows the listed Z_S and Z_L is not the device impedance but rather the impedance that is presented to the device. This is analogous to the

Γ_S and Γ_L that was calculated for the simultaneous conjugate match. The actual device impedance is the conjugate of the given optimum source and load impedance. The designer must be cautious when interpreting the optimum source and load impedances from various vendor data sheets. Some manufacturers may list the actual device impedance as shown on the Freescale Semiconductor data sheet of Figure 7-38. We also need to read the impedance data from the tables rather than directly from the Smith Chart. Because of the very low input and output impedance of power transistors, it is common to normalize the Smith Chart to 10 Ω so that the impedance locus is not compressed on the left hand side of the Smith Chart. The impedances given in the table are the actual impedance rather than the normalized impedance.

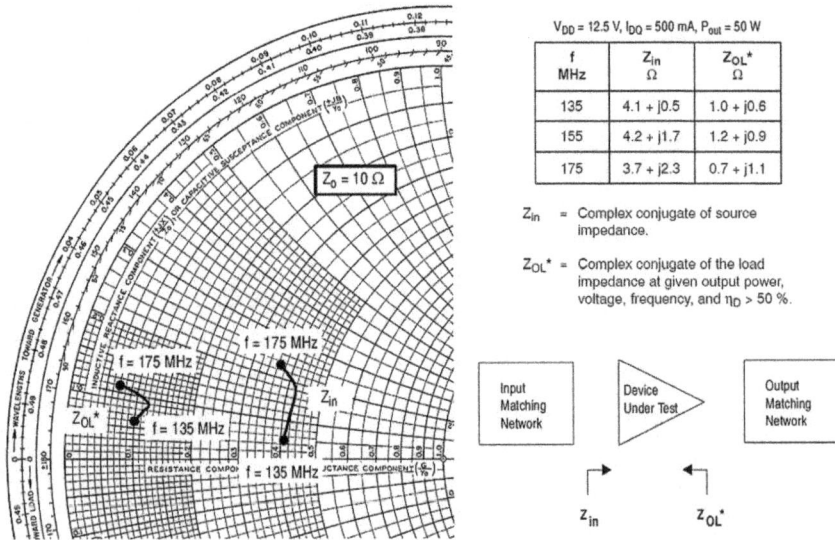

$V_{DD} = 12.5$ V, $I_{DQ} = 500$ mA, $P_{out} = 50$ W

f MHz	Z_{in} Ω	$Z_{OL}*$ Ω
135	4.1 + j0.5	1.0 + j0.6
155	4.2 + j1.7	1.2 + j0.9
175	3.7 + j2.3	0.7 + j1.1

Z_{in} = Complex conjugate of source impedance.

$Z_{OL}*$ = Complex conjugate of the load impedance at given output power, voltage, frequency, and $\eta_D > 50$ %.

Figure 7-38 Large-signal series equivalent impedance (*courtesy of Freescale*)

Power Amplifier Matching Network Design

In this section we will design the matching networks for the NPT25100 GaN power transistor at a frequency of 2.140 GHz. The matching networks of previous examples have all been based on two-element L-networks. In this example we will design three-element Pi network matching circuits. The Pi network is often preferred over a two element network because the component values are less sensitive when physically realizing the impedance match. It can also help to keep the circuit Q lower for improved bandwidth. We will illustrate both graphical techniques and synthesis techniques. The input matching circuit will be designed using graphical techniques while the output matching network will be designed using network synthesis techniques.

Input Matching Network Design

Example 7.7-1 Design a three elements Pi network to match the 50 Ω source impedance to the amplifier input at 2140 MHz.

Solution: Graphical matching techniques is employed for the input matching network using the Smith Chart. We will design the matching circuit moving from 50 Ω at the center to the source impedance as given on the data sheet. A one port S parameter file can be created with impedance data as shown in Figure 7-39. Note the differences in line number 2 of the file. This line defines the format of the data contained within the file. The (Z) defines the file as containing impedance data. The number (1) means that the data is normalized to one, the actual device impedance.

```
!Optimum Source Impedance,Zs, presented to the NPT25100
# GHZ Z RI R 1
!Freq     REZ1            IMZ1
2.140     12.1            -20.0
2.300     10.0            -3.0
2.400     9.5             -3.0
2.500     9.0             -3.0
2.600     8.5             -3.0
2.700     8.0             -3.0
```

Figure 7-39 One port S parameter data file containing the Zs data

Setup a schematic with the one-port Z_S impedance data and sweep it at a fixed frequency of 2140 MHz. Then setup a second schematic with a 50 Ω resistor to represent the center of the Smith Chart and begin the Pi network design. Plot the impedance of both networks on the same Smith Chart. We know that the shunt inductors of a Pi network will travel clockwise on the conductance circles while the series inductor will travel clockwise on a resistance circle. Note the location of the conductance circle in which Z_S is located. The first capacitor and series inductor L section must move to the location of the Z_S conductance circle. Figure 7-40 shows the result.

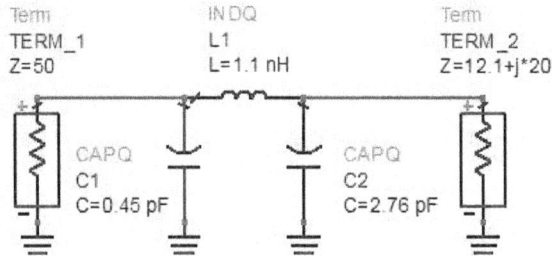

Figure 7-40 Amplifier input matching Pi network

Simulate the schematic and display S11 and S21 in a rectangular plot.

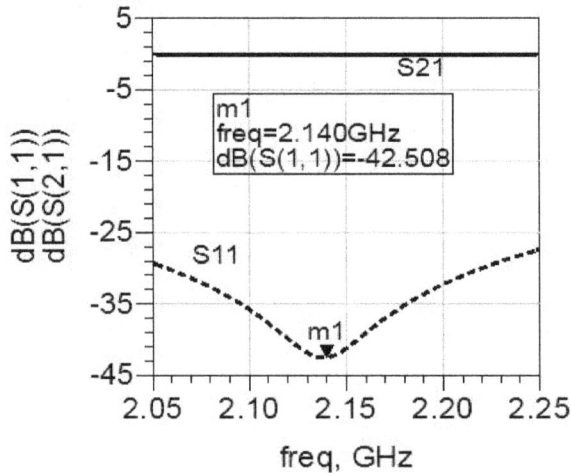

Figure 7-41 Amplifier input matching Pi network

Output Matching Network Design

Example 7.7-2: Design a three elements Pi network to match the 50 Ω load impedance to the amplifier output at 2140 MHz.

Solution: The output matching network will be designed using the one port S parameter file shown in Figure 7-42. Use caution when editing S parameter files on disk to be read into an ADS simulation. The first time that an S parameter file is read from disk it is loaded into the ADS Workspace. Subsequent simulations will not go back to disk to read the file but rather read the file from the Workspace for faster simulation speed. The file will not be read from disk again until the copy loaded into the Workspace has been deleted.

```
!NPT25100 Output Impedance
# GHz Z RI R 1
!Freq      REZ1      IMZ1
2.140      2.6       2.6
2.300      2.5       2.3
2.400      2.5       2.5
2.500      2.5       2.7
2.600      2.5       3.1
2.700      2.5       3.3
```

Figure 7-42 S parameter data file containing the conjugate of Z_L

The Impedance Matching Utility program is used to design the output matching networks. The matching network is shown in Figure 7-43.

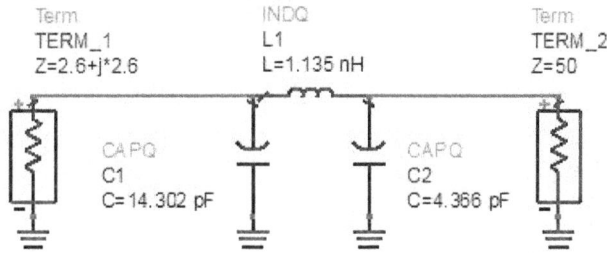

Term
TERM_1
Z=2.6+j*2.6

INDQ
L1
L=1.135 nH

Term
TERM_2
Z=50

CAPQ
C1
C=14.302 pF

CAPQ
C2
C=4.366 pF

Figure 7-43 Output matching network

Simulate the schematic and display S11 and S21 in a rectangular plot.

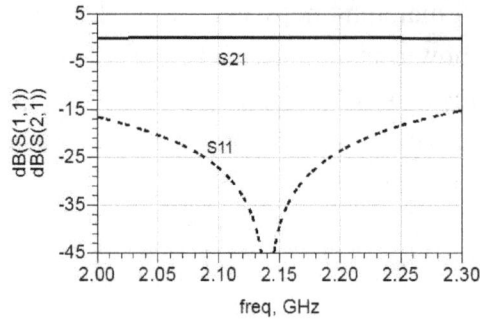

Figure 7-44 Output matching network

Finally a two-port S parameter file can be created that contains the impedance data for the conjugate of Z_S and Z_L. This would be representative of the large signal input and output impedance of the device.

```
!Large Signal Impedances of the NPT25100
# GHz Z RI R 1
!Freq      REZ1   IMZ1      M(S21)      A(S21)      M(S12)      A(S12)      REZ2   IMZ2
 2.140     12.1   20.0      0           0           0           0           2.6    2.5
 2.300     10.0   3.0       0           0           0           0           2.5    2.3
 2.400     9.5    3.0       0           0           0           0           2.5    2.5
 2.500     9.0    3.0       0           0           0           0           2.5    2.7
 2.600     8.5    3.0       0           0           0           0           2.5    3.1
 2.700     8.0    3.0       0           0           0           0           2.5    3.3
```

Figure 7-45 S parameter data file containing the conjugate of Z_S and Z_L

Figure 7-45 shows the resulting matching networks attached to the large signal impedance data for the device. This allows examination of the return loss of the amplifier at the design frequency of 2140 MHz and the usable bandwidth. From Figure 7-45 note that the forward (S21) and reverse (S12) transmission parameters have been set to zero. Because we have no definition of the transmission parameters we cannot evaluate the gain of the circuit. If the manufacturer's data sheet includes an S parameter file along with the large signal impedance data, we could also examine the stability parameters to determine whether the device may require a stabilization network. The linear design techniques do provide a means of performing the matching network design from which the amplifier physical design can be realized. The circuit can then be built and empirically tuned and optimized on the bench for the desired performance. A thorough CAD design of the power amplifier requires a nonlinear physical model for the transistor. Under large signal conditions the strong nonlinearities will cause the gain to change as the drive (input) power changes. The gain of the device will go into compression and decrease as the input power is further increased. Harmonic energy is created by these nonlinearities that will further influence the behavior of the amplifier.

Example 7.7-3: Assemble, simulate, and display the response.

Solution: Create a new schematic in ADS with the input and output matching networks attached to the device. Add a new S parameter simulation that sweeps the amplifier from 2 GHz to 2.2 GHz.

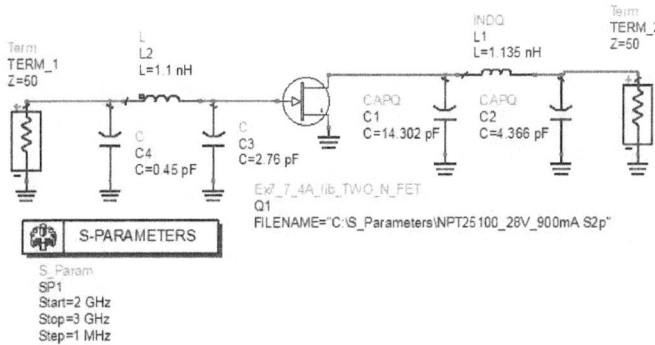

Figure 7-46 Assembly and simulation of the power amplifier

Create a graph and plot the return loss S11 and S22 in a rectangular graph.

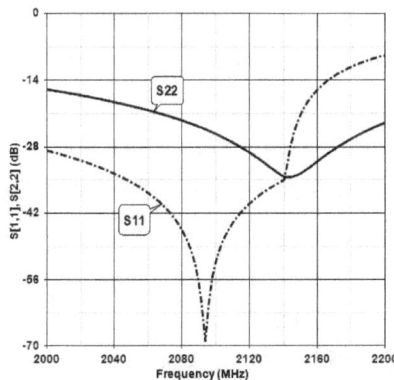

Figure 7-47 Resulting matching networks and return loss

The impedance matching techniques learned in Chapters five and six have been utilized to perform basic linear amplifier matching. These matching techniques have been solved analytically using the equations written in the Matlab script. Graphical techniques have also been applied to the design of multi-element impedance matching using the Smith Chart. These techniques provide the engineer with a comprehensive set of tools to apply to amplifier impedance matching.

References and Further Reading

[1]   Ali A. Behagi, *RF and Microwave Circuit Design,* A Design Approach Using (**ADS**). Techno Search, Ladera Ranch, CA 92694. August 2015.

[2]   Chris Bowick, *RF Circuit Design*, Butterworth-Heinemann, Newton, MA, 1982

[3]   Keysight Technologies, Manuals for Advanced Design Systems, *ADS 2016.01 Documentation Set*, Keysight EEsof EDA Division, Santa Rosa, California   www.keysight.com

[4]   Guillermo Gonzales, *Microwave Transistor Amplifiers – Analysis and Design*, Second Edition, Prentice Hall Inc., Upper Saddle River, NJ.

[5]   Randy Rhea, *The Yin-Yang of Matching: Part 1 – Basic Matching Concepts*, High Frequency Electronics, March 2006

[6]   Steve C. Cripps, *RF Power Amplifiers for Wireless Communications*, Artech House Publishers, Norwood, MA. 1999

[7]   David M. Pozar, *Microwave Engineering*, Third Edition, John Wiley & Sons, New York, 2005

[8]   R. Ludwig, P. Bretchko, *RF Circuit Design*, Theory and Applications, Prentice Hall, Upper Saddle River, NJ, 2000

[9]   Ted Grosch, *Noise Concepts and Design, Noble Publishing*, Atlanta, GA., 2003

Problems

7-1. Design a Maximum Gain Amplifier

(a) For the Agilent HBFP0405 Transistor, calculate the stability factor, K, and the stability measure, B1 as well as Γ_{MS} and Γ_{ML} and conjugate match impedances at 2 GHz.

(b) Use ADS built in functions to determine the same parameters at 2 GHz.

(c) Sweep the frequency range of the device over the entire range of frequencies contained in the S parameter file.

(d) Plot of the stability parameters and GMAX for the Agilent HBFP0405 device from 100 to 4000 MHz.

(e) Employ a parallel RC circuit on the input of the device to stabilize the transistor. Make both the R and C values tunable. Tune the resistance and capacitor values until K > 1 for frequencies above 500 MHz.

(f) Add an RL network in shunt with the input of the transistor for additional low frequency stability. Tune the resistance and inductor values until K > 1 for frequencies above 500 MHz.

(g) Use the analytical impedance matching techniques developed in Chapter 5 to design the input and output matching L-networks. When performing the analytical match we are; matching 50 Ω source impedance to the impedance looking into the device; and 50 Ω load impedance to the impedance looking into output of the device.

(h) Complete the amplifier design in ADS by attaching the input and output matching circuits to the stabilized transistor. Use a linear sweep from 100

MHz to 4000 MHz to analyze the response of the amplifier, S11, S22, and S21 all in dB.

(i) Place a marker on S12 to measure the maximum gain of the amplifier at 2 GHz. Compare the maximum gain with the value originally calculated as G_{MAX}.

7-2. Design a Constant Gain Amplifier

(a) Use the Mitsubishi MGF0911a device to design a specific gain amplifier that achieves 10 dB gain at the frequency of 2 GHz. Examine the stability circles for the device in ADS by creating a schematic with the small signal S parameter file.

(b) Setup a linear analysis with both the start and stop frequencies set at 2 GHz. Add a linear analysis from 100 MHz to 3000 MHz with 100 MHz steps. Then add a Table to display K, B1, and GMAX.

(c) Plot a series of constant power gain circles on the Smith Chart and choose the circle which is the closest to the design goal of 10 dB. The load reflection coefficient, defined as a reflection coefficient, Γ_L, is any point on the circumference of the circle that does not enter the unstable region. Choose a point that is not close to the unstable regions of the stability circles. Read the load impedance visually from the Smith Chart at this point.

(d) Calculate the source impedance from Equation 7-18. The source impedance is therefore defined as a function of the chosen load reflection coefficient.

(e) Use the Smith Chart to graphically design the simple two-element LC input and output matching networks.

(f) Create a new schematic and attach the input and output matching circuits to the device. Create a linear analysis that sweeps the amplifier from 1 GHz to 3 GHz. Display S11 and S21 in a rectangular plot.

(g) Place a marker on S12 to measure the gain of the amplifier at 2 GHz. Compare the gain with the value originally selected as specific gain.

7-3. Design a Low Noise Amplifier

Design a single stage Low Noise Amplifier using Agilent AT-32011 device. The amplifier is intended to operate with a source and load impedance of 50 Ω. The design specifications are given as:

Center Frequency:	500 MHz
Gain:	18 dB minimum
Noise Figure:	1.2 dB maximum
Output Return Loss:	Less than -10 dB

(a) Create an ADS schematic with the device S parameter file. Setup a Linear Analysis and add a Table to display the stability parameters K and B1 and the noise parameters Γ_{opt} and Z_{OPT}. Also add a Smith Chart with Γ_{opt} and the stability circles. Then create a Smith Chart and display the stability circles along with the noise circles.

(b) Drawn the available gain circles are for gains of 0, 1, 2, 3, 4, 5 and 6 dB less than maximum gain.

(c) Utilize the "L" Network Impedance Matching Utility to match the 50 Ω source impedance to Z_{opt} of the device at 500 MHz. In the impedance matching utility you must enter the conjugate of Z_{opt} for the complex load impedance.

(d) Use ADS to calculate the load reflection coefficient, Γ_L, from Equation (7-24). Convert Γ_L to output impedance Z_L. Z_{OUT} is the conjugate of Z_L. Check the location of Γ_L with respect to the output stability circle to make sure that the impedance does not lie in the region of instability. If Γ_L is safely outside of the SB2 circle, use Γ_L as an acceptable output reflection coefficient.

(e) Again use the "L" Network Impedance Matching Synthesis Tool to match the 50 Ohm load impedance to Z_{OUT} of the device at 500 MHz.

(f) Assemble the Low Noise Amplifier and simulate the circuit to see that it meets the design specifications.

Chapter 8

Multi-Stage Amplifier Design

Introduction

Practical amplifiers often involve a cascade of amplifier stages to provide the required gain for a specific application. In an amplifier cascade, the output impedance of one device is typically matched to the input impedance of the succeeding stage rather than a 50 Ω resistance. This chapter provides an overview of amplifier inter-stage matching. For simplicity much of this chapter will focus on ideal schematic elements for inter-stage matching. Finally a brief introduction to cascade analysis is presented as it can be applied to linear simulation in ADS.

Two-Stage Amplifier Design

Figure 8-1 shows a block diagram of a two stage amplifier with the matching networks represented by a box. The input impedance of each device is designated as ZM1 while the output is designated as ZM2. We will revisit two of the transistors used in Chapter 7 to design a two stage amplifier cascade. The first stage is the SHF0189 and the second stage is the RT243. The design specifications are given as:

> Center Frequency = 2350 MHz
> Bandwidth: 2260 MHz to 2380 MHz
> Gain: $\geq$ 34 dB
> Input & Output Return Loss $\leq$ -10 dB

The cascade is designed as a maximum gain-conjugate matched amplifier. Therefore three conjugate matching networks must be designed in accordance with Figure 8-1. The first conjugate matching network is designed between the 50 Ω source and the first stage input impedance, ZM1. The second matching network is the inter-stage matching network.

This network is designed to match the first stage output impedance, ZM2, to the second stage input impedance, ZM1. The third conjugate matching network is designed between the second stage output impedance, ZM2, and the 50 Ω load impedance.

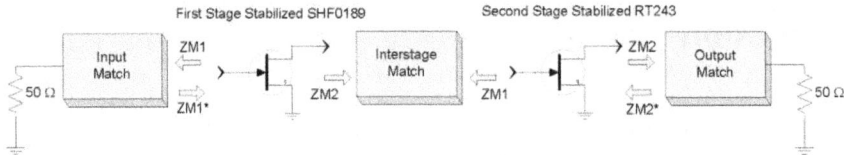

Figure 8-1 Two-stage amplifier with impedance matching networks

Stability Consideration of the First Stage Amplifier Network

The first stage of the amplifier utilizes the SHF-0189 device. We know that the device must be unconditionally stable for the maximum gain conjugate matched amplifier. Using the techniques of Chapter 7 a stabilizing network is designed around the SHF-0189.

Example 8.2-1 Measure the input-output impedance and the maximum stable gain of the first stage of the cascaded amplifier by utilizing the SHF-0189 device at 2350 MHz.

Solution: Once the device is stabilized the maximum stable gain and the required input and output simultaneous match impedance is determined using the built-in functions in ADS. Create a new workspace and a new schematic in ADS. Use the stabilizing network developed in Chapter 7 for the SHF-0189 device as shown in Figure 8-2.

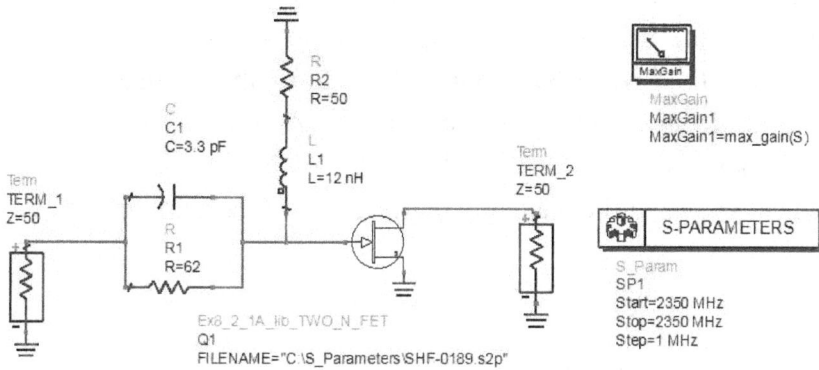

Figure 8-2 Stabilized first stage schematic

Simulate the schematic and display the response. Write the equations for source and load simultaneous match impedances in the data display window, as shown in Figure 8-3.

Eqn Zsource = sm_z1(S)   Eqn ZLoad = sm_z2(S)

freq	MaxGain1	Zsource	ZLoad
2.350 GHz	15.779	6.947 + j22.058	13.773 + j16.471

Figure 8-3 Maximum gain and simultaneous match impedances

As the Table in Figure 8-3 shows, the maximum gain and the input - output impedances are:

$$Zsouce = 6.947 + j22.058 \ \Omega$$
$$Zload = 13.773 + j16.471 \ \Omega$$
$$G_{max} = 15.779 \ dB$$

Amplifier Input Matching Network Design

Example 8.2-2: Design the first stage input matching network.

Solution: When designing the amplifier input matching network, we are matching the 50 Ω source impedance to the conjugate impedance looking into the SHF-0189 device. Therefore, the complex impedance that we are matching to is the conjugate of Zsource. Normalize the load impedance and use Equations (5-32) and (5-33) in Appendix B to calculate the element values of the matching network. The procedure follows.

1. Enter design parameters and normalize the load impedance

Z0=50; RL=6.947; XL=-22.058; f=2.35e9; r=RL/Z0; x=XL/Z0

2. Use the equations in Appendix B to calculate the matching element values

B1=sqrt((1-r)/r)/Z0

X1=Z0*(sqrt(r*(1-r))-x)

L1=X1/(2*pi*f)

C2=B1/(2*pi*f)

The calculation results show that the series element is an inductor L1= 2.665 nH and the shunt element is a capacitor, C2 = 3.372 pF.

To display the response of the matching network, create a new workspace in ADS and open a new schematic window. Insert the S_Params Template and connect the input matching elements as shown in Figure 8-4.

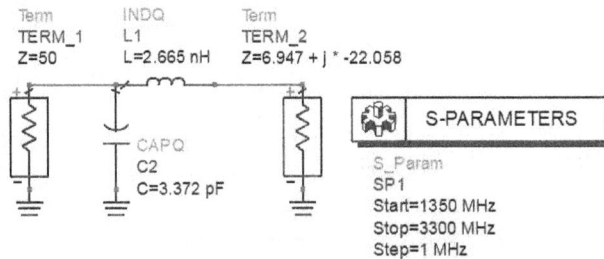

Figure 8-4 Schematic of the input matching network

Simulate the schematic from 1350 MHz to 3350 MHz and display the insertion loss, S21, and return loss, S11, in dB, as shown in Figure 8-5.

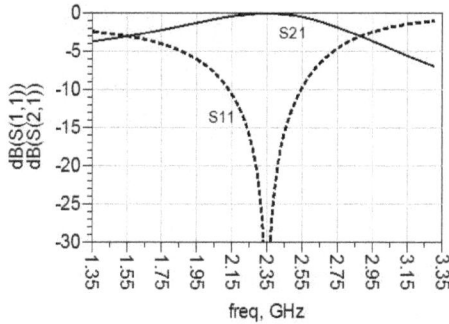

Figure 8-5 Response of the input matching network

Second Stage Impedance Matching Design

Example 8.2-3 Measure the simultaneous match input-output impedance and the maximum stable gain of the second stage of the cascaded amplifier by utilizing the RT243 device at 2350 MHz.

The second stage device along with its stabilization network is shown in Figure 8-6.

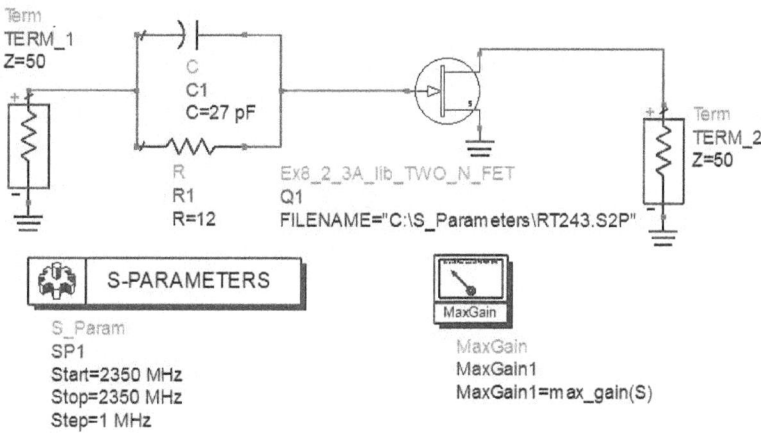

Figure 8-6 Stabilized second stage schematic

As the Figure 8-7 shows, the stabilized RT243 device has a GMAX = 18.973 dB at 2.35 GHz.

Eqn ZLoud=sm_z2(S)   Eqn Zsource=sm_z1(S)

freq	MaxGain1	ZLoud	Zsource
2.350 GHz	18.973	2.094 - j1.780	0.689 - j4.270

Figure 8-7 Stabilized second stage with conjugate match impedance

According to Figure 8-7 the respective impedances maximum gain are given below.

$$Zsource = 0.689 - j4.27 \ \Omega$$
$$Zload = 2.094 - j1.78 \ \Omega$$
$$G_{max} = 18.973 \ dB$$

Inter-Stage Matching Network Design

One common method of inter-stage matching is to design each stage separately into a 50 Ω system and then cascade them directly with no additional matching. Many RF transistors have very low input and output impedance, much less than 50 Ω, making a low impedance interstage network desirable. In cascaded narrowband amplifier design a more efficient method of interstage matching is to conjugately match the two complex impedances directly together with a single L-network and thus reduce the number of interstage matching elements. Therefore, for the interstage matching network, we can use the equations developed in Chapter 5 to directly match together the two complex impedances.

Example 8.2-4: Design the interstage matching network between the SHF-0189 and the RT243 devices.

Solution: The procedure for the solution follows.

1. Enter design parameters and normalize the load impedance

RS=13.773; XS=-16.471; RL=0.688; XL=4.271; f=2.35e9

2. Use the equations in Appendix B to calculate the element values

B3=((RS*XL)+sqrt(RS*RL*(RL^2+XL^2-RS*RL)))/(RS*(RL^2+XL^2))

X3=(RS*XL-RL*XS)/RL+(RL-RS)/(B3*RL)

L2=X3/(2*pi*f)

C1=B3/(2*pi*f)

The calculation results show that the shunt element is an inductor C1=
17.91 pF and the series element is an inductor, L2 = 2.037 nH.

To display the response of the matching network, create a new workspace in
ADS and open a new schematic window. Insert the S_Params Template and
connect the matching elements as shown in Figure 8-8.

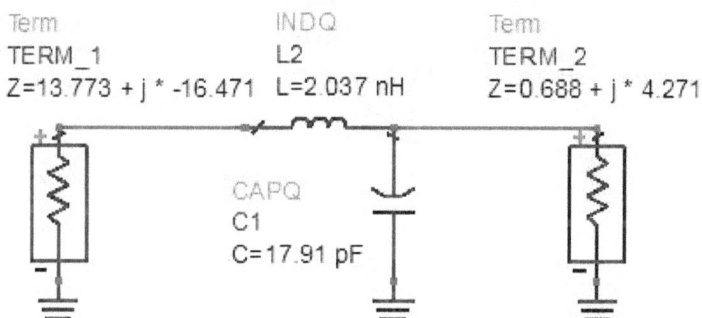

Figure 8-8 Schematic of the interstage matching network

Simulate the schematic from 1350 MHz to 3350 MHz and display both the
return loss, S11, and the insertion loss, S21, in dB, as shown in Figure 8-9.

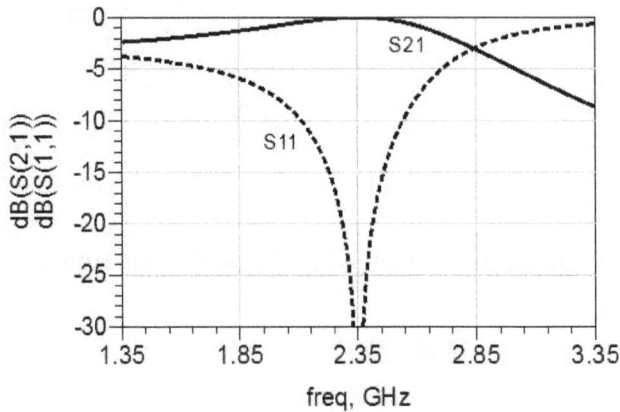

Figure 8-9 Response of the interstage matching network

Second Stage Output Matching Network

Example 8.2-5: Design the second stage output matching network.

Solution: Use Equations (5-24) and (5-25) in the textbook to calculate the element values of the output matching network. The calculation results show that the shunt capacitor C1=6.48 pF and the series inductor L1 = 0.577 nH. Set up a new schematic in ADS and select the element values as shown in Figure 8-10. Simulate the schematic from 1350 MHz to 3350 MHz and display the insertion loss, S21, in dB.

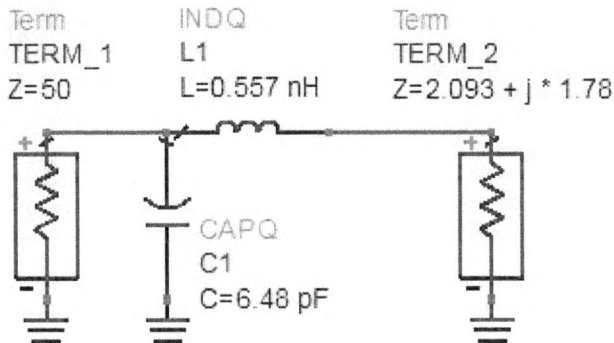

Figure 8-10 Schematic of the output matching network

The simulated response is shown in Figure 8-11.

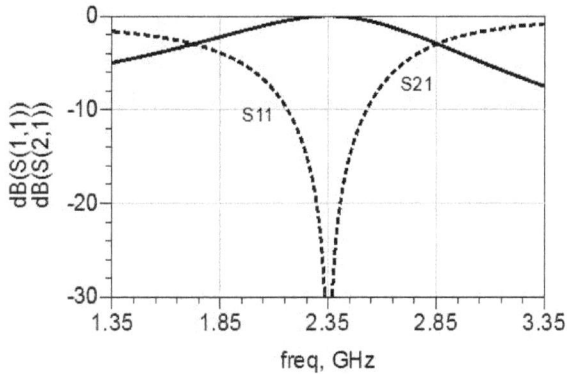

Figure 8-11 Response of the output matching network

Two-Stage Amplifier Simulation

Example 8.3-1: Assemble, simulate, and display the response of the two stage amplifier.

Solution: The matching networks are attached to the stabilized devices to form a two stage amplifier. The input, inter-stage, and output matching networks are cascaded along with the devices as shown in Figure 8-12.
We will only deal with the ideal schematic at this point. The final design, of course, would include the physical models of the lumped element components and the interconnecting microstrip circuit elements.

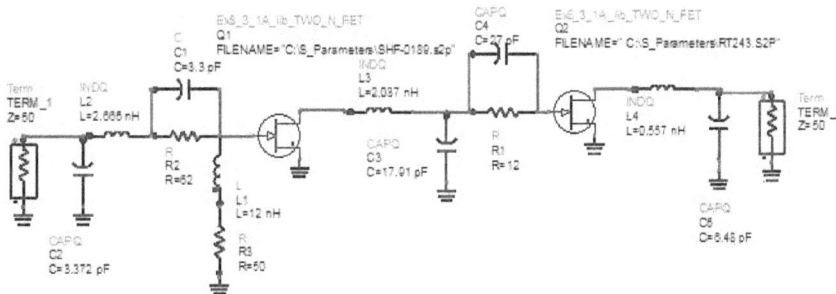

Figure 8-12 Two-stage amplifier schematic with ideal matching networks

The simulated gain and return loss are shown in Figure 8-13. The simulated gain of the two-stage cascade at 2.35 GHz is 34.754 dB which is very close to the direct summation of the GMAX for each stage. The gain of the amplifier is greater than 34 dB from 2260 MHz to 2380 MHz, satisfying one of the design specifications of the amplifier. The input and output return loss is well below -10 dB, thereby satisfying all of the specifications of the amplifier.

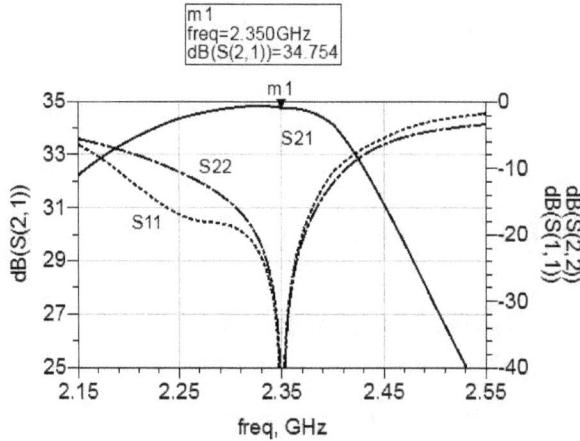

Figure 8-13 Ideal amplifier cascade response

Multi-Stage Low Noise Amplifier Cascade

In practical amplifier applications, a single stage amplifier is rarely adequate to meet the overall gain and output power requirements of an amplifier design. Therefore it is necessary to cascade multiple gain stages to achieve a particular gain specification. In this section the gain and noise figure of the multi-stage cascaded LNA is discussed.

Cascaded Gain and Noise Figure

As we have seen in the example Ex8.3-1, the overall gain of a cascaded amplifier is simply the algebraic sum of the gains (or losses) in dB. The gain of an amplifier cascade is then given by the equation.

$$Cascade\ Gain, G_{dB} = G1_{dB} + G2_{dB} + G3_{dB} + ...$$

But the overall noise figure is defined by the Friis formula.

Solution: As the Friis formula shows, the overall noise figure of a cascade is influenced by all stages in the cascade. The noise figure contribution of the second stage is reduced by the gain of the first stage. Therefore, to maintain a low noise figure it is important that the first stage have high gain. The noise factor of a given stage is reduced by the gain of the preceding stages throughout the cascade.

$$Noise\ Factor,\ F\ =\ F_1 + \frac{F_2 - 1}{G_1} + \frac{F_3 - 1}{G_1 G_2} + \frac{F_4 - 1}{G_1 G_2 G_3} + ...$$

where,

$$Noise\ Factor,\ F = 10^{\frac{F_{dB}}{10}}$$

$$Gain,\ G = 10^{\frac{G_{dB}}{10}}$$

In the solution of the Friis formula it is important to realize that the noise figure and gain must not be in dB format. This is referred to as the noise factor and related to the noise figure by the 10[log(F)] function.

For the three-stage LNA of Figure 8-14, the cascaded gain is easily calculated as 41 dB.

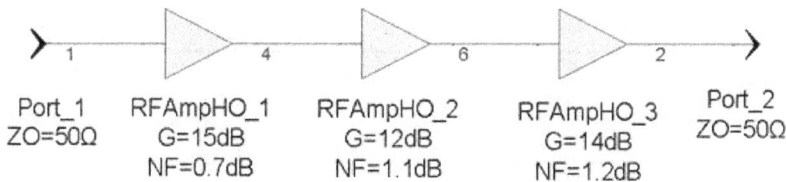

Port_1	RFAmpHO_1	RFAmpHO_2	RFAmpHO_3	Port_2
ZO=50Ω	G=15dB	G=12dB	G=14dB	ZO=50Ω
	NF=0.7dB	NF=1.1dB	NF=1.2dB	

Figure 8-14 Three-stage cascaded low noise amplifier

Using the Friis formula to solve for the resulting noise factor leads to the following result.

$$F = 10^{\frac{0.7}{10}} + \frac{10^{\frac{1.1}{10}} - 1}{10^{\frac{15}{10}}} + \frac{10^{\frac{1.2}{10}} - 1}{10^{\frac{15}{10}} \cdot 10^{\frac{12}{10}}} = 1.1844$$

Converting the noise factor back to noise figure gives a total cascaded noise figure of 0.736 dB.

$$F_{dB} = 10 \cdot \log(1.1844) = 0.736 \ dB$$

The cascaded gain and noise figure can be calculated in ADS using the simple block diagram of Figure 8-14. This three stage cascade uses the RFAmpHO system level model but can be used in linear simulation for gain and noise figure.

Setup a linear simulation for any frequency range as the frequency is independent for this use of the RFAmpHO model. The simulated cascade gain, S21, and noise figure, NF, are shown in Figure 8-15. We can see that the overall noise figure of the cascade is 0.736 dB which correlates with the solution of the Friis formula. The noise figure of the first stage dominates the overall noise figure and is slightly degraded by the 2nd and 3rd stages.

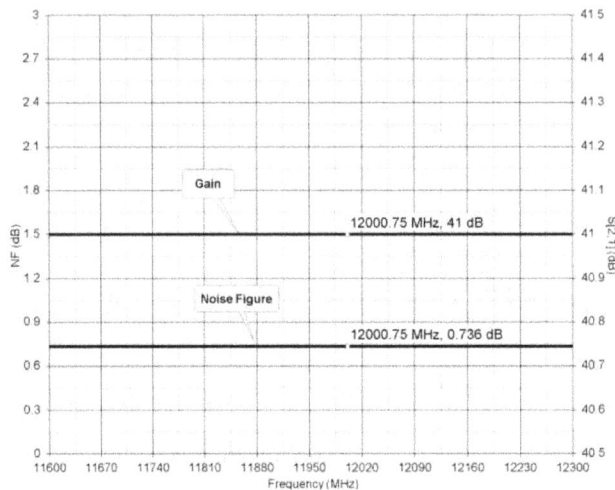

Figure 8-15 Simulated LNA cascade noise figure and gain

Impedance Match and the Friis Formula

A Low Noise Amplifier is typically used as the front end of a radio receiver. Therefore it is often attached to an antenna and filter combination. From Chapter 4 we have seen that the impedance of a filter can vary significantly across its passband as determined by the ripple and return loss. The same can be true of an antenna or the antenna feed network. Because the Noise Figure of an LNA is dependent on its source impedance, large errors can be obtained in the application of the Friis formula when calculating the overall noise figure of such a cascaded network. The Friis formula assumes that there is a perfect impedance match between each stage in the cascade and there are no variations in impedance across the frequency band.

Example 8.4-1A: consider a typical Low Noise Amplifier used in the Ku Band frequency range of 11.7 to 12.2 GHz. To eliminate any interference from the uplink signal or terrestrial sources a low loss waveguide filter is placed at the input of the LNA. This system level block diagram is shown in Figure 8-16.

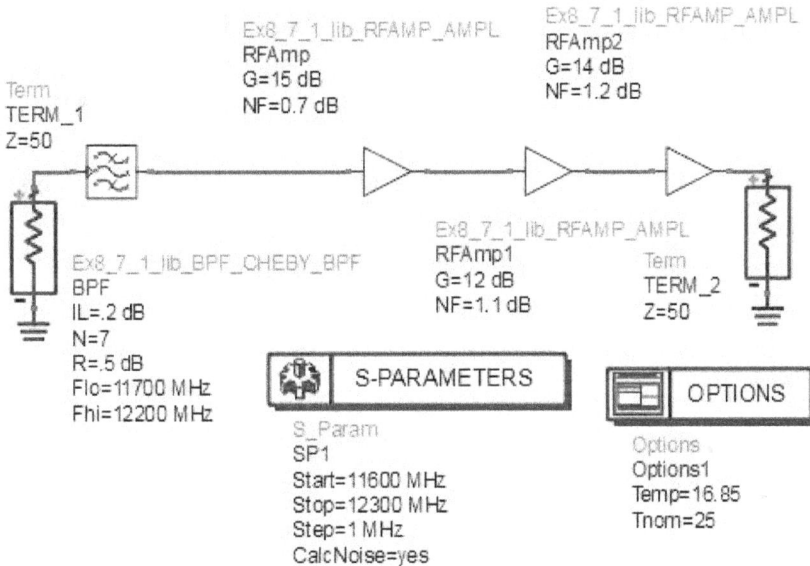

Figure 8-16 Ku band LNA with bandpass filter

Note that a passive device's insertion loss is entered as a negative gain in dB in the Friis formula. The noise figure then becomes the absolute value of this loss or simply 0.2 dB in the case of the bandpass filter. Using the Friis formula to solve for the resulting noise figure leads to the following result.

$$F = 10^{\frac{0.2}{10}} + \frac{10^{\frac{0.7}{10}} - 1}{10^{\frac{-0.2}{10}}} + \frac{10^{\frac{1.1}{10}} - 1}{10^{\frac{-0.2}{10}} \cdot 10^{\frac{15}{10}}} + \frac{10^{\frac{1.2}{10}} - 1}{10^{\frac{-0.2}{10}} \cdot 10^{\frac{15}{10}} \cdot 10^{\frac{12}{10}}} = 1.2402$$

Converting the noise factor back to noise figure gives a total cascaded noise figure of 0.935 dB.

$$F_{dB} = 10 \cdot \log (1.2402) = 0.935 \; dB$$

Based on this simple application of the Friis formula the engineer would expect the system noise figure to be 0.935 dB. Simulating the block diagram of Figure 8-14 in ADS reveals a vastly different result. The simulated noise figure of Figure 8-17 shows that there is significant ripple in the noise figure. The worst case noise figure is actually greater than 2 dB across the passband of the amplifier. The 0.5 dB ripple in the filter actually results in greater than 1.5 dB ripple in the noise figure. This is due to the fact that the noise figure of the input device is very sensitive to the impedance that is presented to it. The 0.935 dB noise figure calculation is an erroneous result that is obtained when using the Friss formula or one of the many spreadsheet cascade analysis programs. By modeling the simple cascade in ADS we can quickly become aware of this condition.

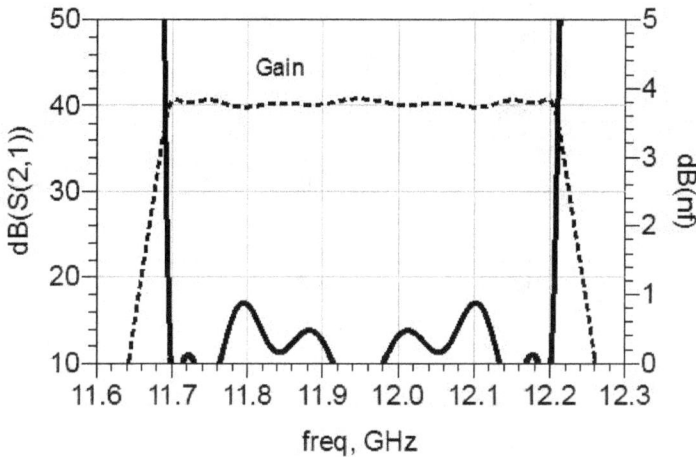

Figure 8-17 Ku band LNA with bandpass filter simulated response

Reducing the Effect of Source Impedance Variation

In practice it is often desirable to place a low-loss isolator at the input of the LNA to buffer the effects of impedance variation due to the filter ripple. It is important that the isolator have very low loss as its insertion loss will also add to the overall noise figure. This can easily be modeled in ADS by adding an isolator to the block diagram as shown in Figure 8-18.

Example 8.4-1B: Add an isolator with an insertion loss of 0.1 dB to the block diagram of Figure 8-16 and plot the LNA cascade insertion loss and noise figure from 11.6 to 12.3 GHz.

Solution: An isolator with 0.1 dB insertion loss reduces the noise figure by 0.8 dB. Perform an S parameter simulation over the range of 11.6 GHz to 112.3 GHz with 600 points. This puts the noise figure of the LNA closer to 1.5 dB, which is a better value for the reception of Ku Band satellite signals from space. The resulting cascaded gain and noise figure is shown in Figure 8-19. Note the decrease in the noise figure due to the addition of the isolator between the filter and the amplifier.

Ex8_7_1_lib_ISO_ISO
Isolator_1
IL=.1 dB
ISO=40 dB

Ex8_7_1_lib_RFAMP_AMPL
RFAmp
G=15 dB
NF=0.7 dB

Ex8_7_1_lib_RFAMP_AMPL
RFAmp2
G=14 dB
NF=1.2 dB

Term
TERM_1
Z=50

Term
TERM_2
Z=50

Ex8_7_1_lib_BPF_CHEBY_BPF
BPF
IL=.2 dB
N=7
R=.5 dB
Flo=11700 MHz
Fhi=12200 MHz

Ex8_7_1_lib_RFAMP_AMPL
RFAmp1
G=12 dB
NF=1.1 dB

S-PARAMETERS

S_Param
SP1
Start=11600 MHz
Stop=12300 MHz
Step=1 MHz
CalcNoise=yes

OPTIONS

Options
Options1
Temp=16.85
Tnom=25

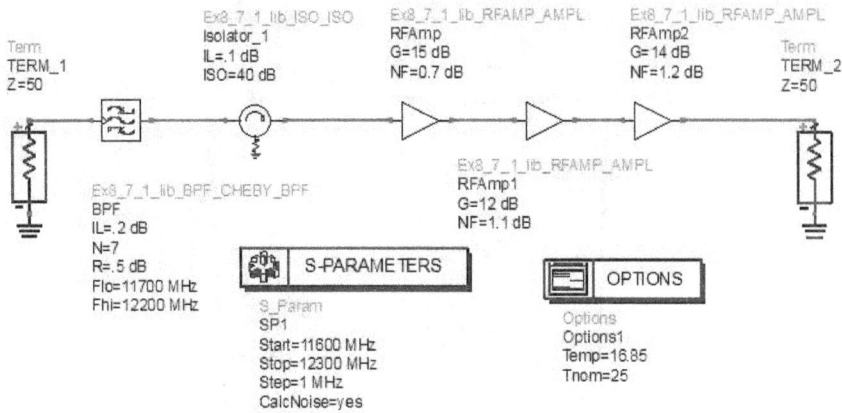

Figure 8-18 Ku band LNA with isolator and bandpass filter

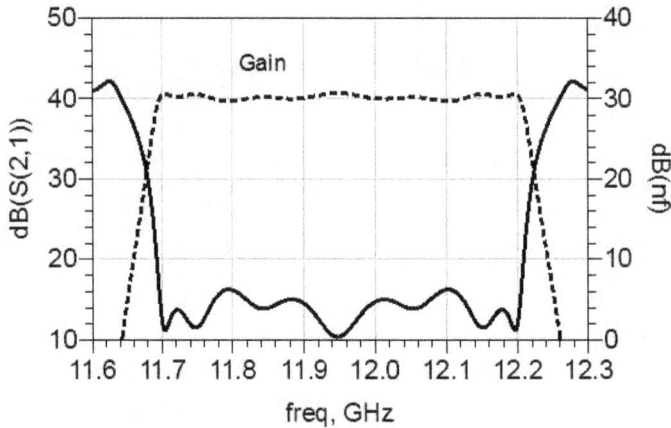

Figure 8-19 LNA cascade noise figure with input matching circuit

Summary

The example of Section 8.4 gives an introduction to the important subject of system, or block diagram, level simulation.

As the previous example shows there are also system level computations that can be evaluated with linear simulation techniques. Linear simulation continues to be a very important topic and is the foundation for all RF and microwave CAD work. This volume provides the reader with a thorough coverage of the linear circuit design techniques that can be accomplished with linear simulation in ADS. As an applications manual this text forms a bridge between the classic theory and practical engineering problem solving.

References and Further Reading

[1]   Ali A. Behagi, *RF and Microwave Circuit Design,* A Design Approach Using (**ADS**). Techno Search, Ladera Ranch, CA 92694. August 2015.

[2]   Randy Rhea, *The Yin-Yang of Matching: Part 1 – Basic Matching Concepts*, High Frequency Electronics, March 2006

[3]   Steve C. Cripps, *RF Power Amplifiers for Wireless Communications*, Artech House Publishers, Norwood, MA. 1999

[4]   David M. Pozar, *Microwave Engineering*, Third Edition, John Wiley & Sons, New York, 2005

[5]   R. Ludwig, P. Bretchko, *RF Circuit Design*, Theory and Applications, Prentice Hall, Upper Saddle River, NJ, 2000

[6]   D. Roddy, J. Coolen, *Electronic Communications*, Second Edition, Reston Publishing Company, Inc., Reston, Virginia, 1981

[7]   Keysight Technologies, Manuals for Advanced Design Systems, *ADS 2016.01 Documentation Set*, Keysight EEsof EDA Division, Santa Rosa, California   www.keysight.com

Problems

8-1. Design a Two–Stage maximum gain amplifier

(a) Use the Agilent HBFP0405 and the Mitsubishi MGF0911a transistors to design a two-stage maximum gain amplifier at 2.2 GHz. All the input, output, and interstage impedance matching networks should be analytically designed using two-element L-networks.

(b) Design the stabilizing network for the first stage using Agilent HBFP0405 transistor

(c) Using the HBFP0405 stabilized transistor create a tabular output of the simultaneous match source and load impedance and Gmax

(d) Design the first stage input matching network by matching the 50 Ohm source impedance to the conjugate impedance looking into the stabilized HBFP0405 device, ZM1.

(e) Design the stabilizing network for the second stage using the Mitsubishi device

(f) Using the MGF0911a stabilized transistor create a tabular output of the simultaneous match source and load impedance and Gmax

(f) Design the second stage output matching network by matching the 50 Ohm load impedance to the conjugate impedance looking into the output of the stabilized MGF0911a device, ZM2

(g) Design the interstage matching network between stage one ZM2 and stage two ZM1.

(h) Cascade the two amplifiers and display the overall response of the cascaded network. The cascaded gain of the amplifier must be the sum of the maximum gains of the two single-stage amplifiers.

8-2. Increase the bandwidth of the matching networks

(a) Increase the bandwidth of the input matching network by selecting a virtual resistor R_M between the input resistor of the matching network, R_S, and the output resistor of the matching network, R_L, such that $R_M = \sqrt{R_S R_L}$ and then conjugately match the input and output impedances to the virtual resistor R_M with L-networks.

(b) Cascade the two matching networks and measure the fractional bandwidth at 20 dB return loss.

(c) Compare the measured fractional bandwidth with the fractional bandwidth of the original matching network at 20 dB return loss and calculate the percentage of the increase in the bandwidth.

(d) Repeat the above procedure for the interstage and output matching networks.

8-3. LNA Cascade Using the Friis Formula

(a) Consider a typical Low Noise Amplifier used in the X Band of 8.7 to 9.2 GHz. Using the Friis formula calculate the overall noise figure of the following LNA cascade.

LNA #1	G1 = 15 dB	F1 = 1.1 dB
LNA #2	G1 = 14 dB	F1 = 1.2 dB
LNA #3	G1 = 13 dB	F1 = 1.3 dB

Appendix A

Straight Wire Parameters for Solid Copper Wire

Wire Size (AWG)	Diameter in Mils	Resistance Ohms/1000 ft.	Area in circular Mils	Suggested Maximum Current Handling, Amperes[1]
0000	460.0	0.049	211600	1000
000	409.6	0.062	167800	839
00	364.8	0.078	133100	665
0	324.9	0.098	105500	527
1	289.3	0.124	83690	418
2	257.6	0.156	66360	332
3	229.4	0.197	52620	263
4	204.3	0.249	41740	208
5	181.9	0.313	33090	165
6	162.0	0.395	26240	131
7	144.3	0.498	20820	104
8	128.5	0.628	16510	83
9	114.4	0.793	13090	65
10	101.9	0.999	10380	52
11	90.7	1.26	8230	41
12	80.8	1.56	6530	32
13	72.0	2.00	5180	26
14	64.1	2.52	4110	20
15	57.1	3.18	3260	16
16	50.8	4.02	2580	13
17	45.3	5.05	2050	10
18	40.3	6.39	1620	8.0
19	35.9	8.05	1290	6.0
20	32.0	10.1	1020	5.0
21	28.5	12.8	812	4.0
22	25.3	16.2	640	3.0
23	22.6	20.3	511	2.5
24	20.1	25.7	404	2.0
25	17.9	32.4	320	1.6
26	15.9	41.0	253	1.2
27	14.2	51.4	202	1.0
28	12.6	65.3	159	0.80
29	11.3	81.2	123	0.61
30	10.0	104.0	100	0.50
31	8.9	131	79.2	0.40
32	8.0	162	64.0	0.32
33	7.1	206	50.4	0.25
34	6.3	261	39.7	0.19
35	5.6	331	31.4	0.16
36	5.0	415	25.0	0.12
37	4.5	512	20.2	0.10
38	4.0	648	16.0	0.08
39	3.5	847	12.2	0.06
40	3.1	1080	9.61	0.05
41	2.8	1320	7.84	0.04
42	2.5	1660	6.25	0.03
43	2.2	2140	4.84	0.024
44	2.0	2590	4.00	0.020
45	1.76	3350	3.10	0.016
46	1.57	4210	2.46	0.012
47	1.40	5290	1.96	0.010
48	1.24	6750	1.54	0.008
49	1.11	8420	1.23	0.006
50	0.99	10600	0.98	0.005

Appendix B

B-1 Impedance Matching Network Design

One of the important tasks in RF and microwave engineering is the determination of how an arbitrary complex load impedance, $Z_L = R_L + jX_L$, is analytically matched to any complex source impedance, $Z_S = R_S + jX_S$. This problem arises mainly in the design of inter-stage matching networks between active devices or between an antenna and a transmitter where both impedances are usually complex.

In Figure B-1 the complex load impedance, $Z_L = R_L + j\,X_L$, is to be matched to the source impedance $Z_S = R_S + jX_S$. The only condition for impedance matching is that both R_S and R_L must be nonnegative while X_S and X_L could take any real value. In the *RF and Microwave Circuit Design* textbook it was proved that the maximum power is transferred from the source to the load when the load impedance is the conjugate of the source impedance.

In designing impedance matching networks using only 2 lumped elements, such as L-networks, there are two configurations that can match an arbitrary load impedance to any arbitrary source impedance. In the first configuration the first element adjacent to the load is a series element and the second element is a shunt element. Such configuration is shown in Figure B-1.

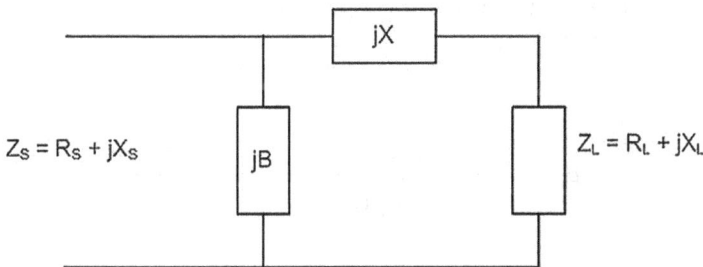

Figure B-1 First impedance matching network configuration

In the second configuration the first element adjacent to the load is a shunt element and the second element is a series element, as shown in Figure B-2.

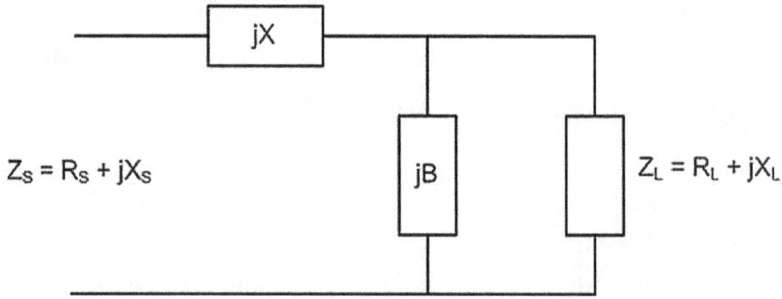

Figure B-2 Second impedance matching network configuration

B-2 Matching a Complex Load to a Complex Source Impedance

To match a complex load to any complex source impedance using a single L-network, either the first or the second configuration may be used. The choice of configurations depends on the conditions that source and load impedances dictate. Applying the maximum power transfer condition $Z_S^* = Z_{IN}$ to the first matching configuration in Figure B-2 we get:

$$R_S - jX_S = \frac{1}{jB + \left(\dfrac{1}{jX + R_L + jX_L} \right)} \qquad (5\text{-}6)$$

By separating the real and imaginary parts of Equation (5-6) we obtain two solutions for B and X as follows:

$$B_1 = \frac{R_L X_S + \sqrt{R_L R_S (R_S^2 + X_S^2 - R_L R_S)}}{R_L (R_S^2 + X_S^2)} \qquad (5\text{-}7)$$

$$X_1 = \frac{R_L X_S - R_S X_L}{R_S} + \frac{R_S - R_L}{B_1 R_S} \qquad (5\text{-}8)$$

And

$$B_2 = \frac{R_L X_S - \sqrt{R_L R_S (R_S^2 + X_S^2 - R_L R_S)}}{R_L (R_S^2 + X_S^2)} \qquad (5\text{-}9)$$

$$X_2 = \frac{R_L X_S - R_S X_L}{R_S} + \frac{R_S - R_L}{B_2 R_S} \qquad (5\text{-}10)$$

Similarly, applying the same procedure to the second configuration we have:

$$R_S - jX_S = jX + \frac{1}{jB + \left(\dfrac{1}{R_L + jX_L} \right)} \qquad (5\text{-}11)$$

Separating the real and imaginary parts of Equation (5-11), we also get two sets of solutions for B and X:

$$B_3 = \frac{R_S X_L + \sqrt{R_L R_S (R_L^2 + X_L^2 - R_L R_S)}}{R_S (R_L^2 + X_L^2)} \qquad (5\text{-}12)$$

$$X_3 = \frac{R_S X_L - R_L X_S}{R_L} + \frac{R_L - R_S}{B_3 R_L} \qquad (5\text{-}13)$$

And

$$B_4 = \frac{R_S X_L - \sqrt{R_L R_S (R_L^2 + X_L^2 - R_L R_S)}}{R_S (R_L^2 + X_L^2)} \qquad (5\text{-}14)$$

$$X_4 = \frac{R_S X_L - R_L X_S}{R_L} + \frac{R_L - R_S}{B_4 R_L} \qquad (5\text{-}15)$$

Conditions for the validity of solutions are that the arguments of the square roots in Equations (5-7), (5-9), (5-12) and (5-14) be positive or zero.

Therefore:

- If $R_S^2 + X_S^2 - R_L R_S > 0$ and $R_L^2 + X_L^2 - R_L R_S < 0$, the two solutions in Equations (5-7) through (5-10) are the only valid solutions.

- If $R_S^2 + X_S^2 - R_L R_S < 0$ and $R_L^2 + X_L^2 - R_L R_S > 0$, the two solutions in Equations (5-12) through (5-15) are the only valid solutions.

- If $R_S^2 + X_S^2 - R_L R_S > 0$ and $R_L^2 + X_L^2 - R_L R_S > 0$, all four solutions in Equations (5-7) through (5-10) and (5-12) through (5-15) are valid.

Once the real values for B and X are calculated, the values of the matching elements are obtained from the following equations:

If B is positive, the matching element is a capacitor given by:

$$C = \frac{B}{2\pi f} \qquad (5\text{-}16)$$

If B is negative, the matching element is an inductor given by:

$$L = -\frac{1}{2\pi f B} \qquad (5\text{-}17)$$

If X is positive, the matching element is an inductor given by:

$$L = \frac{X}{2\pi f} \qquad (5\text{-}18)$$

If X is negative, the matching element is a capacitor given by:

$$C = -\frac{1}{2\pi f X} \qquad (5\text{-}19)$$

In the above equations, if the frequency is in Hz, capacitor and inductor values are in Farad and Henry, respectively.

In practice, we utilize Equations (5-7) through (5-15) to design the matching L-networks.

B-3 Matching Complex Load to Real Source Impedance

In single stage amplifier design a common matching problem is the matching of a complex load impedance, $Z_L = R_L + jX_L$ to real source impedance, $Z_S = R_S$. The complex impedance is usually the load and the real impedance is the characteristic impedance of the transmission line connected to the source. For the case of real source impedance the matching configuration of Figure B-1 is redrawn in Figure B-3.

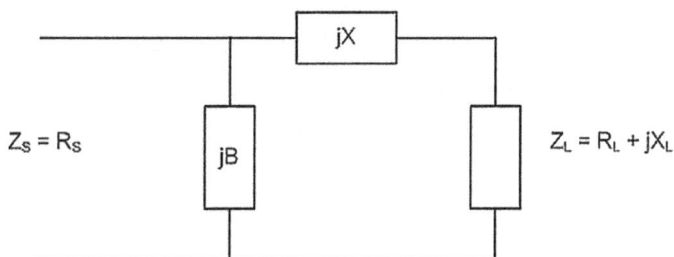

Figure B-3 Matching a resistive source to a complex load impedance

To derive the analytical expressions for B and X we utilize the maximum power transfer condition and set the conjugate of the source impedance equal to input impedance of the matching network followed by the load impedance, as given in Equation (5-20).

$$R_S = \cfrac{1}{jB + \left(\cfrac{1}{jX + R_L + jX_L}\right)} \tag{5-20}$$

Note that the conjugate of a real source resistor is equal to itself.

The solutions for B and X in Equation (5-20) can easily be obtained by substituting $X_S = 0$ in Equations (5-7) through (5-10).

$$B_1 = \frac{+\sqrt{R_S - R_L}}{R_S \sqrt{R_L}}$$

(5-21)

$$X_1 = +\sqrt{R_L(R_S - R_L)} - X_L$$

(5-22)

And,

$$B_2 = \frac{-\sqrt{R_S - R_L}}{R_S \sqrt{R_L}}$$

(5-23)

$$X_2 = -\sqrt{R_L(R_S - R_L)} - X_L$$

(5-24)

Note that the solutions given in Equations (5-21) through (5-24) are only valid if $R_L < R_S$. To calculate B and X, when $R_L > R_S$, we use the second matching configuration shown in Figure B-4,

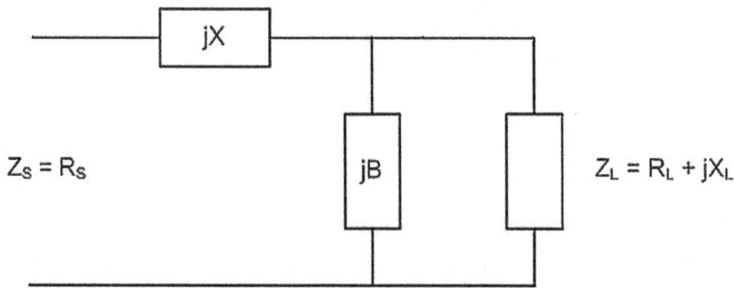

Figure B-4 Matching complex load to resistive source (2nd configuration)

Applying the maximum power condition we have:

$$R_S = jX + \cfrac{1}{jB + \left(\cfrac{1}{R_L + jX_L}\right)}$$

(5-25)

The solutions for B and X in Equation (5-25) can be obtained by reusing Equations (5-12) and (5-14) and substituting $X_S = 0$ in Equations (5-13) and (5-15).

$$B_3 = \frac{R_S X_L + \sqrt{R_L R_S (R_L^2 + X_L^2 - R_L R_S)}}{R_S (R_L^2 + X_L^2)} \tag{5-26}$$

$$X_3 = \frac{R_S X_L}{R_L} + \frac{R_L - R_S}{B_3 R_L} \tag{5-27}$$

And

$$B_4 = \frac{R_S X_L - \sqrt{R_L R_S (R_L^2 + X_L^2 - R_L R_S)}}{R_S (R_L^2 + X_L^2)} \tag{5-28}$$

$$X_4 = \frac{R_S X_L}{R_L} + \frac{R_L - R_S}{B_4 R_L} \tag{5-29}$$

The conditions for the valid solutions are that the arguments of the square roots in Equations (5-26) through (5-29) be non-negative. Therefore, the two solutions obtained from Equations (5-26) through (5-29) are valid only if $R_L > R_S$. The combined conditions are summarized in Table B-1.

Case #	First Condition	Second Condition	# of Solutions	Equations Used
1	$R_L < R_S$	$R_L^2 + X_L^2 - R_L R_S > 0$	4	(5-21) to (5-24) (5-26) to (5-29)
2	$R_L < R_S$	$R_L^2 + X_L^2 - R_L R_S < 0$	2	(5-21) to (5-24)
3	$R_L > R_S$	N/A	2	(5-26) to (5-29)

Table B-1 Impedance matching conditions and the number of solutions

The solutions in Equations (5-26) through (5-29) can be simplified by normalizing the load impedance with respect to the source resistor. Therefore, if $R_S = Z_0$, the normalized load resistance is:

$$r = \frac{R_L}{Z_0} \tag{5-30}$$

And

$$x = \frac{X_L}{Z_0} \tag{5-31}$$

The simplified equations for matching a complex load to a real source are:

$$B_1 = \frac{\sqrt{\dfrac{(1-r)}{r}}}{Z_0} \tag{5-32}$$

$$X_1 = Z_0 \left[\sqrt{r(1-r)} - x \right] \tag{5-33}$$

$$B_2 = -\frac{\sqrt{(1-r)}}{Z_0} \tag{5-34}$$

$$X_2 = -Z_0 \left[\sqrt{r(1-r)} + x \right] \tag{5-35}$$

$$B_3 = \frac{x + \sqrt{r(r^2 + x^2 - r)}}{Z_0 (r^2 + x^2)} \tag{5-36}$$

$$X_3 = Z_0 \sqrt{\frac{(r^2 + x^2 - r)}{r}} \tag{5-37}$$

$$B_4 = \frac{x - \sqrt{r\left(r^2 + x^2 - r\right)}}{Z_0\left(r^2 + x^2\right)}$$ (5-38)

$$X_4 = -Z_0\sqrt{\frac{\left(r^2 + x^2 - r\right)}{r}}$$ (5-39)

B-4 Matching a Real Load to a Real Source Impedance

When source and load impedances are both real, the first matching configuration can be applied.

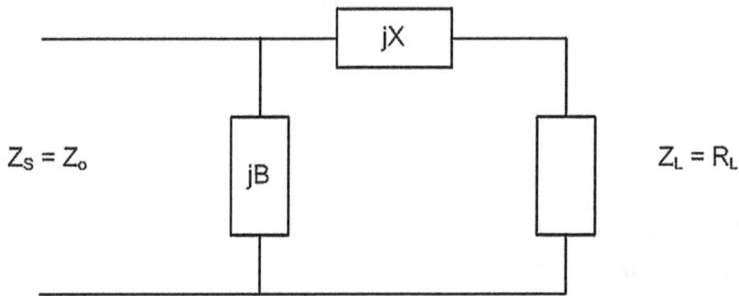

Figure B-5 First matching configuration with $X_L = X_S = 0$

To design the matching network, apply the maximum power transfer condition and require that $Z_0^* = Z_{IN}$. Therefore,

$$Z_0 = \frac{1}{jB + \left(\dfrac{1}{jX + R_L}\right)}$$ (5-40)

Substituting $r = \dfrac{R_L}{Z_0}$ in Equation (5-40), we get:

$$Z_0 = \frac{1}{jB + \left(\dfrac{1}{jX + rZ_0} \right)} \tag{5-41}$$

The solutions for B and X can be obtained by using Equations (5-32) and (5-34) and substituting x = 0 in Equations (5-33) and (5-35).

$$B_1 = \frac{\sqrt{(1-r)/r}}{Z_0} \tag{5-42}$$

$$X_1 = Z_0 \sqrt{r(1-r)} \tag{5-43}$$

$$B_2 = -\frac{\sqrt{(1-r)/r}}{Z_0} \tag{5-44}$$

$$X_2 = -Z_0 \sqrt{r(1-r)} \tag{5-45}$$

Note that the two solutions given by Equations (5-42) through (5-45) are only valid if r is less than 1 or $R_L < Z_0$.

If r > 1, we use the second matching configuration as in Figure B-6.

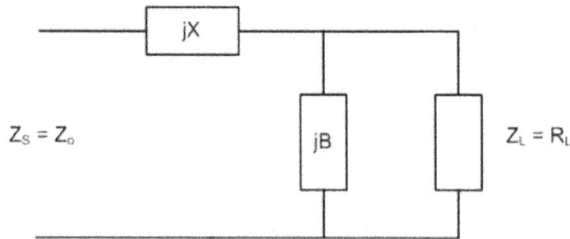

Figure B-6 Second matching configuration for resistive load and source

To calculate the B and X values, we require that, $Z_0{}^* = Z_{IN}$, therefore,

$$Z_0 = jX + \cfrac{1}{jB + \left(\cfrac{1}{Z_0 r} \right)}$$

(5-46)

The solutions for B and X in Equation (5-46) can be obtained by substituting x = 0 in Equations (5-36) through (5-39).

$$B_3 = \frac{\sqrt{r-1}}{Z_0 r}$$

(5-47)

$$X_3 = Z_0 \sqrt{r-1}$$

(5-48)

$$B_4 = \frac{-\sqrt{r-1}}{Z_0 r}$$

(5-49)

$$X_4 = -Z_0 \sqrt{r-1}$$

(5-50)

Note that the two solutions given by Equations (5-47) through (5-50) are only valid if r is greater than 1 or $R_L > Z_0$.

The combined conditions for the validity of solutions are given below.

Case 1 If r is less than 1 there are two L-networks that match the two impedances. The two solutions are given by Equations (5-42) through (5-45).

Case 2 If r is greater than 1 there are two L-networks that match the two impedances. The two solutions are given by Equations (5-47) through (5-50).

Appendix C

C-1 Analytical Design of the Quarter-Wave Matching Networks

We have shown that the input impedance of a quarter-wave network with characteristic impedance, Z_0, terminated in a resistive load, R_L, is given by:

$$Z_{IN} = \frac{Z_0^{\,2}}{R_L}$$

This equation can be written as:

$$Z_O = \sqrt{R_L Z_{IN}} \qquad (6\text{-}1)$$

We have also shown that the maximum power transfer from a source with resistance R_S to a network terminated in a resistive load R_L is achieved only if the input impedance of the network is equal to the source resistance.

$$Z_{IN} = R_S$$

Therefore, for maximum power transfer, the characteristic impedance of the quarter-wave network must satisfy the Equation (6-2).

$$Z_O = \sqrt{R_S R_L} \qquad (6\text{-}2)$$

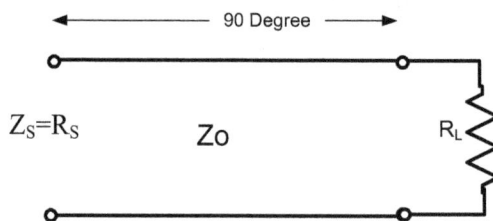

Figure C-1 Quarter-wave network terminated in resistor R_L

Equation (6-2) states that the characteristic impedance of the quarter-wave network, matching R_L to R_S, must be equal to the square root of the product of source and load resistors.

If we normalize the load resistor R_L with respect to R_S,

$$r = \frac{R_L}{R_S} \tag{6-3}$$

Equation (6-2) can be written as Equation (6-4).

$$Z_O = R_S \sqrt{r} \tag{6-4}$$

Notice that for $R_L > R_S$, the characteristic impedance Z_0 is greater than R_S while for $R_L < R_S$, the characteristic impedance Z_0 is less than R_S.

The fractional bandwidth of a network, *FBW*, is defined in Equation (6-5) where f_H and f_L are the upper and lower frequencies of the bandwidth and the center frequency f_0 is equal to $\sqrt{f_H f_L}$, respectively.

$$FBW = \frac{f_H - f_L}{\sqrt{f_H f_L}} \tag{6-5}$$

The fractional bandwidth of a quarter-wave matching network is given in Equation (6-6):

$$FBW = 2 - \frac{4}{\pi} \cdot \cos^{-1}\left(\frac{2\Gamma_m \sqrt{r}}{\sqrt{1 - \Gamma_m^2} \, |1 - r|} \right) \tag{6-6}$$

Where, Γ_m is the magnitude of the reflection coefficient.

Equation (6-6) shows that the fractional bandwidth of a quarter-wave matching network depends on the magnitude of the input reflection coefficient, Γ_m, and the mismatch ratio, r. The solution to Equation (6-6) is only valid if,

$$\frac{2\Gamma_m \sqrt{r}}{\sqrt{1-\Gamma_m^2}\,|1-r|} \leq 1$$

At 3 dB return loss the reflection coefficient is $\Gamma_m = 0.707$, therefore, Equation (6-6) reduces to:

$$FBW_{3dB} = 2 - \frac{4}{\pi}\cdot\cos^{-1}\left(\frac{2\sqrt{r}}{|1-r|}\right) \qquad (6\text{-}7)$$

At 3 dB return loss Equation (6-7) has valid solutions only if,

$$\frac{2\sqrt{r}}{|1-r|} \leq 1 \quad or \quad r^2 - 6r + 1 \geq 0 \qquad (6\text{-}8)$$

Similarly, if we define the bandwidth at $\Gamma_m = 0.1$, corresponding to 20 dB return loss, as a good matching bandwidth it is insightful to evaluate the fractional bandwidth associated with this 20 dB return loss. From Equation (6-6) the fractional bandwidth at $\Gamma_{in} = 0.1$, is:

$$FBW_{20dB} = 2 - \frac{4}{\pi}\cdot\cos^{-1}\left(\frac{0.2\sqrt{r}}{\sqrt{0.99}\,|1-r|}\right) \qquad (6\text{-}9)$$

At 20 dB return loss Equation (6-9) has valid solutions only if,

$$\frac{0.2\sqrt{r}}{\sqrt{0.99}\,|1-r|} \leq 1 \quad or \quad 99r^2 - 202r + 99 \geq 0 \qquad (6\text{-}10)$$

The loaded quality factor, Q_L, of the quarter-wave matching network is defined as the inverse of the fractional bandwidth at 3 dB return loss; therefore, the loaded Q factor can be calculated from Equation (6-11).

$$Q_L = \frac{1}{FBW_{3dB}} = \frac{1}{2 - \frac{4}{\pi} \cdot \cos^{-1}\left(\frac{2\sqrt{r}}{|1-r|}\right)} \qquad (6\text{-}11)$$

Equation (6-11) shows that the validity condition in Equation (6-8) for the 3 dB fractional bandwidth is the same for the Q factor except that whenever the 3 dB fractional bandwidth tends towards infinity the Q factor tends towards zero. The Q factor given in Equation (6-11) is not to be confused with the unloaded transmission line Q factor. This is actually an external Q factor as it relates to the loaded Q of the overall network.

If f is the center (design) frequency, the bandwidth of the circuit is then calculated from Equation (6-12)

.

$$BW = (f) \cdot (FBW) \qquad (6\text{-}12)$$

C-2 Analytical Design of the Series Transmission Line

It has been shown in the *RF and Microwave Circuit Design* textbook that the input impedance of a lossless transmission line of length d and characteristic impedance Z_0 terminated in an arbitrary load Z_L, is given in the following equation.

$$Z_{IN} = Z_o \frac{Z_L + jZ_o \tan \beta d}{Z_o + jZ_L \tan \beta d}$$

Setting $Z_0 = R_S$ and $\tan \beta d = t$, the input admittance of the network in Figure C-1 can be written as:

$$Y_{IN} = \frac{1}{Z_{IN}} = \frac{R_S + jZ_L t}{R_S(Z_L + jR_S t)} = G_{IN} + jB_{IN}$$

(6-13)

Substituting the normalized load impedance, $Z_L/R_s = r + jx$ into Equation (6-13) and separating its real and imaginary parts we get:

$$G_{IN} = \frac{r(1+t^2)}{R_S(r^2 + x^2 + t^2 + 2xt)}$$

(6-14)

And,

$$B_{IN} = \frac{xt^2 + (r^2 + x^2 - 1)t + x}{R_S(r^2 + x^2 + t^2 + 2xt)}$$

(6-15)

The value of d, which implies t, can be obtained by setting the input conductance, G_{IN}, equal to source conductance:

$$\frac{r(1+t^2)}{R_S(r^2 + x^2 + t^2 + 2xt)} = \frac{1}{R_S}$$

(6-16)

Equation (6-16) can be rearranged as:

$$(r-1)\cdot t^2 - 2xt - (r^2 + x^2 - r) = 0$$

(6-17)

Notice that the quadratic Equation (6-17) has two solutions for t. For a wider bandwidth and lower loss usually the smaller value of t is selected. The two solutions for t are Equations (6-180 and (6-19):

$$t_1 = \frac{x + \sqrt{r(r^2 + x^2 - 2r + 1)}}{r - 1}$$

(6-18)

And,

$$t_2 = \frac{x - \sqrt{r(r^2 + x^2 - 2r + 1)}}{r - 1} \qquad (6\text{-}19)$$

With $\tan \beta d = t$, and $\beta \lambda = 2\pi$, we have $d = \dfrac{\lambda}{2\pi} \tan^{-1} t$ and the two solutions for d are:

$$d_1 = \frac{\lambda}{2\pi} \tan^{-1} t_1 \qquad t_1 \geq 0 \qquad (6\text{-}20)$$

$$d_2 = \frac{\lambda}{2\pi} \tan^{-1} t_2 \qquad t_2 \geq 0 \qquad (6\text{-}21)$$

To specify the lengths of d_1 and d_2 in electrical degrees, we get:

$$d_1 = \frac{360}{2\pi} \tan^{-1} t_1 \qquad t_1 \geq 0 \qquad (6\text{-}22)$$

$$d_2 = \frac{360}{2\pi} \tan^{-1} t_2 \qquad t_2 \geq 0 \qquad (6\text{-}23)$$

Because at every half wavelength the input impedance of a transmission line repeats, there are an infinite number of transmission line lengths that matches the load to source impedance. Usually the shorter length is selected to improve the matching bandwidth. If t_1 or t_2 is negative, we add half a wavelength to each line to get positive d_1 and d_2.

$$d_1 = \frac{360(\pi + \tan^{-1} t_1)}{2\pi} \qquad t_1 < 0 \qquad (6\text{-}24)$$

$$d_2 = \frac{360(\pi + \tan^{-1} t_2)}{2\pi} \qquad t_2 < 0 \qquad (6\text{-}25)$$

C-3 Analytical Design of Shunt Stub Matching Networks

To calculate the electrical length of the shunt stub, first substitute t_1 and t_2 in Equation (6-15) to determine B_1 and B_2.

$$B_1 = \frac{xt_1^2 + (r^2 + x^2 - 1)t_1 + x}{R_S(r^2 + x^2 + t_1^2 + 2xt_1)} \tag{6-26}$$

$$B_2 = \frac{xt_2^2 + (r^2 + x^2 - 1)t_2 + x}{R_S(r^2 + x^2 + t_2^2 + 2xt_2)} \tag{6-27}$$

Then, the electrical lengths of the open circuited stubs are found by setting the susceptance of the stubs equal to the negative of the input susceptance.

$$so_1 = \frac{-\lambda\left(tan^{-1}(R_S B_1)\right)}{2\pi} \tag{6-28}$$

$$so_2 = \frac{-\lambda\left(tan^{-1}(R_S B_2)\right)}{2\pi} \tag{6-29}$$

If either stub length in Equations (6-28) or (6-29) is negative, add one half wavelength to obtain a positive stub length. For short-circuited stubs, the two solutions are:

$$ss_1 = \frac{\lambda\left(tan^{-1}\left(\frac{1}{R_S B_1}\right)\right)}{2\pi} \tag{6-30}$$

$$ss_2 = \frac{\lambda\left(tan^{-1}\left(\frac{1}{R_S B_2}\right)\right)}{2\pi} \tag{6-31}$$

About the Author

Ali A. Behagi received the Ph.D. degree in electrical engineering from the University of Southern California and the MS degree in electrical engineering from the University of Michigan. He has several years of industrial experience with Hughes Aircraft and Beckman Instruments. Dr. Behagi joined Penn State University as an associate professor of electrical engineering in 1986. He has devoted over 20 years to teaching microwave engineering courses and directing university research projects. While at Penn State he received the National Science Foundation grant, to establish a microwave and RF engineering lab, and the Agilent ADS software grant to use in teaching high frequency circuit design courses and laboratory experiments. After retirement from Penn State he has been active as an educational consultant. Dr. Behagi is a Keysight Certified Expert, a Senior Member of the Institute of Electrical and Electronics Engineers (IEEE), and the Microwave Theory and Techniques Society.